Virtual Reality and the Built Environment

Virtual Reality and the Built Environment

Jennifer Whyte

Architectural Press
OXFORD AMSTERDAM BOSTON LONDON NEW YORK PARIS
SAN DIEGO SAN FRANCISCO SINGAPORE SYDNEY TOKYO

Architectural Press
An imprint of Elsevier Science
Linacre House, Jordan Hill, Oxford OX2 8DP
225 Wildwood Avenue, Woburn MA 01801-2041

First published 2002

British Library Cataloguing in Publication Data
Whyte, Jennifer
Virtual reality and the built environment
1. Virtual reality in architecture
I. Title
720.2'856

Library of Congress Cataloguing in Publication Data
A catalogue record for this book is available from the Library of Congress

ISBN 0 7506 5372 8

For information on all Architectural Press publications visit our website at www.architecturalpress.com

Composition by Scribe Design, Gillingham, Kent, UK
Printed and bound in Great Britain

Contents

Foreword by Professor David Gann

The way in which we visualize buildings – their component parts, how they work and how they might be used – has a strong bearing on the built environment we create and inhabit. Emerging tools for design visualization are changing the practice of design itself. They provide opportunities, as designers no longer need to be temporally and spatially constrained by previous limitations of sequential decision-making processes. They make it possible to create virtual prototypes, to model attributes and to simulate performance characteristics without having to build full-scale mock-ups. By adding another dimension to the ways in which space can be configured over time, they complement and enhance the value of using face-to-face communications and physical models.

This book provides a rich insight into the development and use of virtual reality – a new tool for design, production and management of the built environment. It shows how changes are occurring; what they mean for professionals in the project team and supply chain; what they mean for clients, managers and end-users; and how new design technologies can be managed in future. It does so by drawing upon case studies from leading users and examples of different practices from around the world.

The book sheds new light on the topic because of the way in which it engages with the process of technological change, within the context of design practice. It shows how virtual reality only became technically possible through developments in a number of underpinning, generic technologies – rapid computing, visualization screens and large databases, together with high speed communications infrastructure. The integration of these technologies has opened new possibilities for applications across the spectrum of design, production and management activities.

As this book shows, rather than leading to uniform processes and standard design practices, these tools are being used in many divergent ways across different segments of the design community. There are expectations of further technological refinement and cost reduction, and this is likely to stimulate more widespread use in future. This book provides a thought-provoking and practical guide to how design organizations – large and small – might benefit by engaging with these new technologies of design. It illustrates the excitement of designing in a multimedia environment and creates a real sense of how we might integrate different parts of the processes of design, production and management to provide better buildings.

David Gann
Programme on Innovation in the Built Environment
SPRU – Science and Technology Policy Research
January 2002

Preface

Virtual reality is influencing the way that spaces are designed and it is changing our experience of the built environment. For example, in the summer of 2000, the artist Horst Kiechle was using a computer for design. Later that year, the spaces he designed were fabricated and installed in a gallery in Sydney. The exhibition, which was entitled *Northwestwind Mild Turbulence*, was enjoyed by visitors to the gallery and by many other people who experienced it through a virtual reality (VR) model.

This book is for professionals, such as architects, engineers and planners, as well as for students and others interested in buildings and cities. The central question it addresses is how virtual reality can be used in the design, production and management of the built environment. We take a fresh look at applications of virtual reality in the construction sector with the aim of inspiring and informing future use.

Virtual reality applications are based on a range of technologies evolved for entertainment, military and advanced manufacturing purposes. As with other emerging technologies, realizing the early dreams for virtual reality has taken longer than was initially predicted (Brooks, 1999). Potential benefits, such as its use by engineering organizations to simulate dynamic operation and co-ordinate detail design, have not always been anticipated. Our understanding of the relative importance of technologies has changed over time. For example, head-mounted displays are less widely used than predicted in the late 1980s. Yet, whilst these symbols of early virtual reality seem increasingly dated, the interactive, spatial, real-time medium at the heart of VR applications is becoming ubiquitous.

Underlying the book is a belief that we can learn from the leading industrial users of virtual reality. Many case studies are included, which are based on interviews with practitioners across the construction sector and in other leading sectors. The book asks many questions. It asks how professionals within the project team – architects, engineers, construction managers, etc. – can benefit from using virtual reality. It also asks how others, such as clients, facility managers and end-users, can benefit from wider involvement and how planners can use virtual reality at the urban scale.

The book considers three key questions. What are the business drivers for the use of virtual reality? What are its limitations? How can virtual reality be implemented within organizations? Leading organizations that use virtual reality have found many different answers to these questions. From the growing pool of industrial examples, I have tried to pick case studies that best illustrate particular positions and that are of lasting interest, rather than simply those that use the most up-to-date technologies. Whilst some good examples will have been missed, I hope that enough are included to give readers a flavour of the business drivers for, and issues related to, the use of virtual reality in design, production and management of the built environment.

A broad definition of virtual reality is taken in this book. As well as high-end immersive VR systems, there are many low-end interactive 3D systems, evolved from the same families of technologies. These are being widely used in industry and are making interactive, spatial, real-time applications available on desktop and mobile computing devices. Including interactive 3D systems in the definition of virtual reality, Frampton (2001) estimates that the worldwide market for VR systems is worth US$348 million in 2001. Many books on virtual reality exclusively describe high-end systems, focusing on hardware and software and only speculating as to its use. In contrast this book focuses on the practical applications rather than platforms and technologies *per se*.

Acknowledgements

This book would never have been completed without the good will of a very large number of people. I would particularly like to thank James Soutter, David Gann, Ammon Salter, Martin Whyte, Dino Bouchlaghem and Tony Thorpe for their comments on earlier drafts of this book and the research on which it is based. I would also like to thank the editorial staff at Architectural Press, Katherine MacInnes and Alison Yates, for their patience and encouragement. The fruits of the labour of many professionals are described in this book. They contributed through participation in case studies and I would like to thank them for sharing their insights and examples with me. Among those that I am particularly indebted to are Johan Bettum, Bruce Cahan, Rennie Chadwick, Steven Feiner, Martin Fischer, Roger Frampton, Lars Hesselgren, Bill Jepson, Carl Johnson, Sawada Kazuya, Scott Kerr, Michael Kwartler, Sebastian Messer, Joan Mitchell, Ken Millbanks, John Mould, Susan O'Leary, Brian O'Toole, Kimon Onuma, Steve Parnell, Alan Penn, Matthew Pilgrim, Shawn Priddle, Hani Rashid, Mervyn Richards, Benedict Schwegler, David Throssell, Lukardis von Studnitz, Hugh Whitehead and Jonathon Zucker.

All quotations that are not otherwise attributed have been taken from transcripts of interviews with professionals. Every effort has been made to check details with all relevant organizations and to ensure that all the appropriate permissions have been obtained. The final text and opinions expressed within it are my own. I bear responsibility for any errors and omissions and will seek to rectify errors at the earliest possible date.

Jennifer Whyte
Programme on Innovation in the Built Environment
SPRU – Science and Technology Policy Research
January 2002

Picture credits

I would like to acknowledge the help of many organizations and individuals who kindly allowed me to include images of their work. Considerable effort has been made to obtain accurate information about these images and the correct wording for crediting the sources as well as copyright permissions. The author and publishers apologize for any errors and omissions and, if notified, will endeavour to correct these at the earliest available opportunity. Images are copyright © and courtesy of the organizations and individuals credited below.

Figure 1.1: Matsushita Electric Works, Japan – reproduced from Sawada (2001).
Figure 1.2: Superscape PLC interactive 3D technology. http://www.superscape.com/
Figure 1.4: Activeworlds.com, Inc.
Figure 1.5: Fakespace Systems Inc.
Figure 1.6: MENSI, provided by AG Electro-Optics Ltd. http://www.ageo.co.uk/laser_scanning/
Figure 1.7: Andy Smith, Centre for Advanced Spatial Analysis (CASA), UCL, London, UK.
Figure 1.8: Vassilis Bourdakis and CASA, University of Bath, UK.
Figures 2.4 and 2.5: Theatron Ltd.
Figures 2.6 and 2.7: Johan Bettum, Norway.
Figure 2.8: out of copyright, but reproduced from 1931 OS map with the kind permission of the Ordnance Survey.
Figure 2.10: Asymptote Architecture.
Figure 2.11: oosterhuis.nl, Noord-Holland Pavilion Version 4.1. Project architect: Kas Oosterhuis. Design team: Kas Oosterhuis, Sander Boer, Ilona Lénárd, Yael Brosilovski, Petra Frimmel, Natasa Ribic. All scripts/3dmodels/renderings by oosterhuis.nl.
Figure 2.12: oosterhuis.nl, www.trans-ports.com, project designer: Kas Oosterhuis. Design team 1999–2001:

Kas Oosterhuis, Andre Houdart, Ilona Lénárd, Ole Bouman, Nathan Lavertue, Philippe Müller, Richard Porcher, Franca de Jonge, Leo Donkersloot, Birte Steffan, Jan Heijting, Arthur Schwimmer, Chris Kievid, Michi Tomaselli, Michael Bittermann, Hans Hubers. All scripts/3dmodels/renderings by oosterhuis.nl.
Figure 2.13: Perilith. http://www.perilith.com/
Figure 2.14: Parallel Graphics. http://www.parallelgraphics.com/
Figure 2.15: Electronic Visualization Laboratory, University of Illinois at Chicago, USA.
Figure 2.16: Parallel Graphics. http://www.parallelgraphics.com/
Figures 3.1 and 3.2: Bechtel – Advanced Visualization/Virtual Reality, San Francisco, CA, USA.
Figures 3.3 and 3.4: WS Atkins – reproduced from Woods (2000) and Kerr (2000).
Figures 3.5–3.7: Virtual Presence Ltd.
Figure 3.8: SHELL – MENSI, provided by AG Electro-Optics Ltd – http://www.ageo.co.uk/laser_scanning/
Figure 3.9: images generated by NavisWorks. NavisWorks is a registered trademark of NavisWorks Ltd., Sheffield, UK.
Figures 4.1 and 4.2: Roderick Lawrence – reproduced from Lawrence (1987).
Figure 4.3: Matsushita Electric Works – reproduced from Sawada (2001).
Figure 4.4: BMW AG and Realtime Technology AG, Germany.
Figure 4.5–4.8: Bechtel – Advanced Visualization/ Virtual Reality, San Francisco, USA.
Figure 4.9: Antycip UK.
Figures 4.10 and 4.11: Phillippe Van Nedervelde, E-SPACES, Germany.
Figures 5.1–5.7: Micheal Kwartler, Environmental Simulation Center, Ltd.
Figures 5.8 and 5.9: Artemedia AG, Germany.
Figures 5.10 and 5.11: Evans and Sutherland. http://www.es.com/ Source material courtesy of the City and County of Honolulu Department of Planning and Permitting.
Figures 5.12 and 5.13: CAD CENTER Corp., Japan.
Figures 6.1 and 6.2: Bechtel – Advanced Visualization/ Virtual Reality, San Francisco, CA, USA.
Figures 6.3 and 6.4: Viasys Oy, Finland.

Plate 1: S. Feiner, B. MacIntyre, M. Haupt and E. Solomon, Columbia University, NY, USA – reproduced from Feiner *et al.* (1993).
Plate 2: Bill Jepson, Urban Simulation Team, UCLA School of Arts and Architecture, CA, USA.
Plate 3: Horst Kiechle, Sydney VisLab, Australia.
Plate 4: Boston Dynamics and the Institute of Creative Technology (ICT).
Plates 5 and 6: Arcus Software. http://www.arcussoft.com/
Plate 7: Mott MacDonald – images created using STEPS software tool.
Plate 8: MultiGen Paradigm. http://www.multigen.com/
Plates 9–12: Laing Construction – Plates 11 and 12 are generated from NavisWorks. NavisWorks is a registered trademark of NavisWorks Ltd., Sheffield, UK.
Plates 13–16: Bechtel London Visual Technology Group.
Plates 18 and 19 (and cover illustration): oosterhuis.nl, www.trans-ports.com, project designer: Kas Oosterhuis. Design team 1999–2001: Kas Oosterhuis, Andre Houdart, Ilona Lénárd, Ole Bouman, Nathan Lavertue, Philippe Müller, Richard Porcher, Franca de Jonge, Leo Donkersloot, Birte Steffan, Jan Heijting, Arthur Schwimmer, Chris Kievid, Michi Tomaselli, Michael Bittermann, Hans Hubers. All scripts/3dmodels/renderings by oosterhuis.nl.
Plates 20 and 21: Mirage 3D and architects Prent Landman, Holland.
Plate 22: Bill Jepson, Urban Simulation Team, UCLA School of Arts and Architecture, USA.
Plate 23: Urban Data Solutions, Inc. http://www.u-data.com/
Plates 24 and 25: Evans and Sutherland. http://www.es.com/. Plate 24. Source material courtesy of the City and County of Honolulu Department of Planning and Permitting. Plate 25. Source material courtesy Aspen Resource Consultants.
Plates 26 and 27: Artemedia AG, Germany.
Plates 28–30: Skyscraper Digital, a division of Little and Associates Architects, Charlotte, NC, USA.

1 Using virtual reality

How much of what we hear is hype? Virtual reality has been widely discussed, but how can it be useful to professionals and others? Can its use improve the quality of the built environment? Can its use improve user involvement? There is no substitute for experience, and this book presents the experience of leading practitioners. We explore the business benefits of and barriers to the use of virtual reality.

Researchers have argued that everyone can use virtual reality, that it is a generic technology that may form an interface to all construction applications. Not all of the leading practitioners share this vision. Virtual reality is being used in industry for a range of different tasks. Some see its use as a specialist activity and, as yet, no company is using it across all functions. Virtual reality is most widely used at the later stages of design, but there is not one single approach to its use. Instead there is a set of related strategies, drivers and models.

However, patterns of use are emerging and some commonalities exist. For example, it is striking that organizations implementing and using virtual reality make a major distinction between models created for professional uses within the project team and supply chain, and those for wider interactions:

1. *within the project team and supply chain*, models are being created and used by consultant engineers, contractors, sub-contractors and suppliers. They may be used internally within one organization or in conjunction with other professional organizations involved in the same project; and
2. *outside the project team*, models are being used for wider interactions with end-users, clients, managers, funding institutions and planners. These models may be quite different from those used by professionals working on the project.

There are different priorities for creating and using models for these two purposes. In later chapters we will explore these uses of virtual reality. We will also look at how the same data is used and reused in models for both purposes.

The book builds on a series of interviews with leading practitioners. First, in this chapter we look at what virtual reality is, how it has developed and how virtual reality models are created. In Chapter 2 we will look at how virtual reality and other forms of representation are different from reality, and how these differences may be used in different tasks to illuminate hidden structure.

In Chapter 3 we explore the use of virtual reality within the project team and supply chain, for the engineering design of complex buildings. In Chapter 4 we look at its use to support design and wider involvement. In Chapter 5 we explore the use of virtual reality for professional purposes and for wider interactions in planning and management at the urban scale. In the final chapter, Chapter 6, the arguments are summarized and we look at the use of virtual reality within the organization.

As virtual reality is a dynamic medium, it cannot be fully represented in still images. Please refer to the book's Website http://www.buildingvr.com for links to related resources and many of the online models mentioned in the book.

What is virtual reality?

The term 'virtual reality' (VR) was first used in the 1980s. The Oxford English Dictionary (OED) points to this early use:

> Virtual reality is not a computer. We are speaking about a technology that uses computerized clothing to synthesize a shared reality. (OED, 1989)

Use of the term has shifted as underlying technologies have become more established. In a report by the US National Research Council (NRC) the following examples are given:

> Simple VR systems include home video games that produce three-dimensional (3D) graphical displays and stereo sound and are controlled by an operator using a joystick or computer keyboard. More sophisticated

> systems – such as those used for pilot training and immersive entertainment experiences – can include head-mounted displays or large projection screens for displaying images, 3D sound, and treadmills that allow operators to walk through the virtual environment. (NRC, 1999: box 10.1)

The term 'virtual reality' has become used to describe applications in which we can interact with spatial data in real-time. It is a buzzword around which communities of industrial users, suppliers, governments, funding bodies and academics have gathered. Other words describe the same or overlapping groups of technologies. These include: 'virtual environments', 'visualization', 'interactive 3D (i3D)', 'digital prototypes', 'simulation', 'urban simulation', 'visual simulation' and '4D-CAD'.

Use of the term 'virtual reality' can direct attention to either the VR medium or the VR system. When the term is used to refer to the VR medium there is a focus on the virtual environment and the model created within the computer. In contrast, when it is used to refer to the VR system the focus is on the hardware and software.

Virtual reality medium

McLuhan explains that 'the "message" of any medium or technology is the change of scale or pace or pattern that it introduces into human affairs' (McLuhan, 1964: 8).

Considering virtual reality as a medium, our attention is focused on the representations within the medium and their implications, rather than the hardware and software of current computer systems. Our interest is how people use and can use virtual reality in the design, production and management of the built environment. As a medium, virtual reality has three defining characteristics. It is:

1 *interactive* – users can interact with models;
2 *spatial* – models are represented in three spatial dimensions; and
3 *real-time* – feedback from actions is given without noticeable pause.

The extent to which these defining characteristics are present may vary. For example, the nature and extent of interaction varies according to the application. Users of virtual reality can normally navigate freely through models, and make decisions about what to look at. However, they may or may not be able to intuitively create objects within the virtual environment. They may or may not be able to change the parameters of objects and change the conditions in which they are viewed. It is some degree of interaction that distinguishes virtual reality from animations and walkthroughs and some minimum interaction is required for a medium to be considered as virtual reality.

The minimum definition of virtual reality as medium – interactive, spatial and real-time – covers a range of applications on different types of VR systems. It includes both the professional applications for construction scheduling and the applications for use at the customer interface. Professionals that use virtual reality emphasize how the medium enables them to understand real-world data about the built environment. One VR supplier put it:

> The majority of our customers have the need to visualize something that is really there, or something that is to be built in context of what is really there, and be able to interact with it ... walk/fly/drive through the scene ... without any constraint.

Many in the construction sector associate virtual reality with the peripherals – head-mounted displays, haptic gloves and joysticks – that were used in early demonstrations. Yet, leading industrial users of virtual reality stress the relationship between the visualization and the engineering and design data. The aim in using virtual reality as a medium is to better understand the built environment as a product and to gain insight into the processes of its construction and operation.

Virtual reality systems

Virtual reality systems support the use of an interactive, spatial, real-time medium and are comprised of the computer hardware and software, the input and output devices, the data and the users. These systems are classified as immersive, non-immersive or augmented reality:

- *Immersive* systems totally surround the user, supposedly providing an unmediated experience. They do this through specialist hardware such as head-mounted and large wall-mounted displays. They require high-end computing power to provide a high realism environment.
- *Non-immersive* systems typically use more generic hardware. The same software techniques are used but the system does not totally immerse the viewer.

1.1
An immersive high-end system – the image shows a built environment application viewed on the immersive display at Matsushita Electric Works in Japan

Sometimes described as window-on-a-world systems, they allow the user to see virtual reality through a screen or display that does not take up their total field of view.

- *Augmented reality* systems overlay virtual and real world imagery allowing the user to interact with both the virtual and real world, for example through the use of mixed video and computer images. Such systems reduce the amount of geometry that it is necessary to build in the virtual world (Plate 1).

There is a spectrum of different types of systems, from high-end immersive systems to low-cost non-immersive systems. This spectrum is polarized with many high-end and many low-end systems. High-end VR systems are designed to give the users a sense of presence: i.e., a sense of 'being there' in a mediated environment

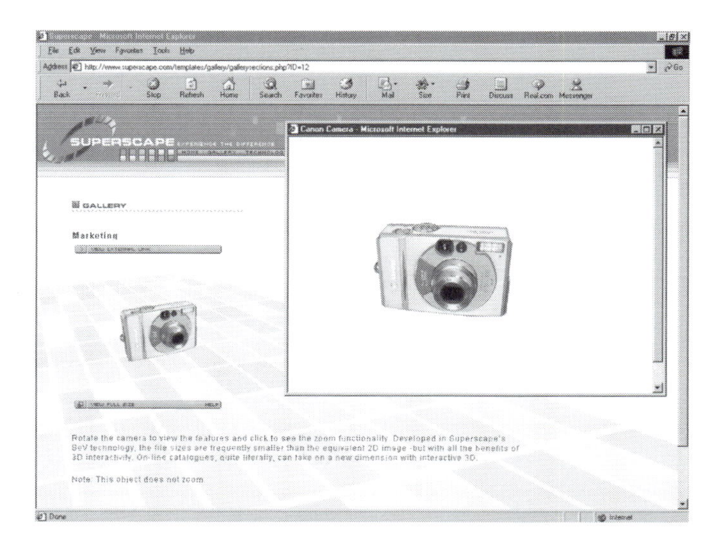

1.2
A non-immersive or window-on-a-world system – the image shows interactive 3D technology being used to showcase products online. Here Superscape's interactive technology is being used to allow a camera and a CD Walkman to be viewed

(Ijsselsteijn *et al.*, 2000). The immersion and presence that they provide may be important for some built environment applications. Some people argue that they are necessary for true virtual reality (Gigante, 1993).

There is increasing interest in augmented and mixed real/virtual applications where the user can be simultaneously looking at virtual data and aware of their real world context, rather than being completely immersed. This is being explored in various construction-related university laboratories, and in corporate research and development (R&D) departments, such as that of the consultant engineering company ARUP (Pilgrim *et al.*, 2001).

The components of a VR system are the computer hardware and software, the input and output devices, the data and the users, as shown in Figure 1.3.

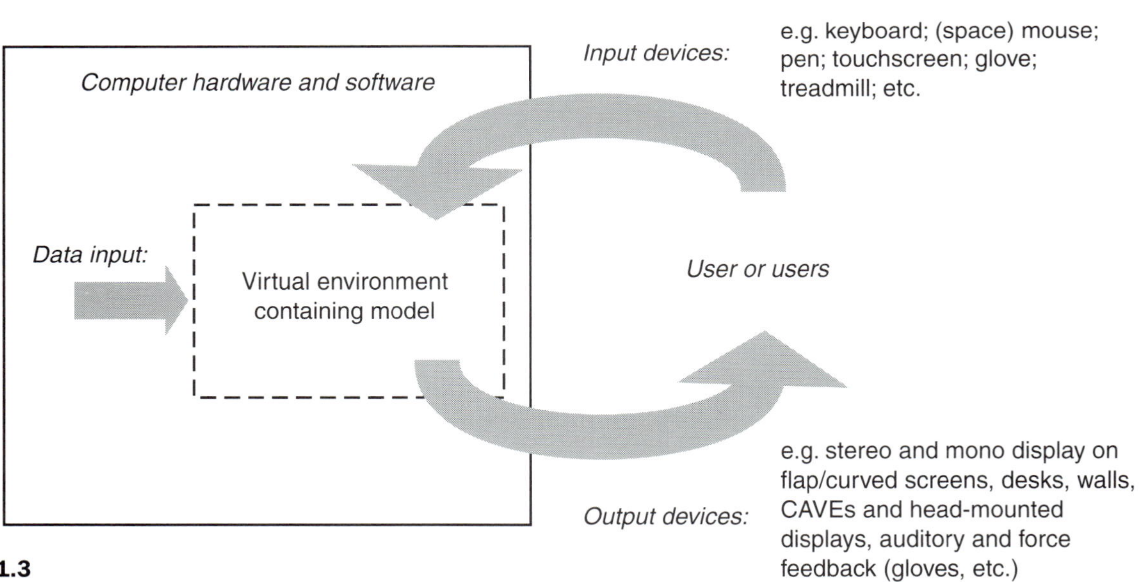

1.3
Components of a VR system – hardware and software, the input and output devices, the data and the users

Though virtual reality is historically associated with high-end computing, a wide range of hardware and software is being used in VR systems. As computers become more ubiquitous, this range increases with interactive 3D (i3D) being used on desktop personal computers (PCs) and on mobile computing devices.

Peripheral input and output devices can be used to make interaction with virtual environments more intuitive. These include methods for position tracking devices, allowing head and eye movements of users to be tracked, and control devices as well as visual, aural and haptic input and feedback (Isdale, 1998).

- *Position tracking and control* – the simplest control hardware is a conventional mouse, trackball or joystick. Though position tracking should ideally include three measures for position (X, Y, Z) and three measures for orientation (roll, pitch, yaw), these devices do not allow this. Use of ultrasonic, magnetic and optical position trackers has been explored to enable six degree of freedom position tracking and control in high-end systems.
- *Visual* – experienced through sight, visual displays of virtual environments can be stereoscopic, with a different picture viewed through each eye, or monoscopic, with both eyes seeing the same picture. Immersive visual displays include the head-mounted display, whilst non-immersive displays include the desktop monitor and workbench.
- *Aural* – experienced through hearing, aural inputs and outputs are often neglected in the industrial use of virtual reality. Yet Brooks (1999) describes how audio quality may be more important than visual quality in some applications.
- *Haptic* – experienced through touch and force. Brooks (1999) is convinced that much of the sense of presence and participation in vehicle simulators comes from the fact that the near-field haptics are exactly right. It is possible to reach out and touch on the simulator everything reachable on the real vehicle.

A key part of the VR system is the data. Models may be built within the virtual environment, but are more usually imported from CAD. They can also be obtained directly from the physical world using techniques such as 3D laser scanning, photogrammetry or geometry capture from film. The users that interact with the data can also be seen as integral to the system.

Historical context

The historical context within which virtual reality has been developed affects our understanding of it. It shapes the way we approach the use of virtual reality as a medium and as a system.

Development of the virtual reality medium

Virtual reality is changing the pace of human affairs but it is not doing this in isolation. It can be seen in the context of longer-term historical trends. The development of high-quality glass in the fourteenth century has led to the world increasingly being viewed through a frame (Mumford, 1934). This frame has made it possible to see certain elements of reality more clearly and has focused attention on a sharply determined and bounded field of view (Foster and Meech, 1995). The development of glass also encouraged later innovations, such as lenses and mirrors that further affected the way we view the world and ourselves.

The coincidence of developments in lens and mirror technologies and the development of accurate portraiture at around 1420 suggests that the link between ways of seeing and the technologies of visualization is much older than usually described (Hockney, 2001). Many of the great masters of Western Art from that time on, such as Caravaggio, Vermeer, Velázquez, van Eyck, van Dyck, etc., may have used lenses in the process of making images (Hockney, 2001). The engineer of the cathedral in Florence, Brunelleschi, would have had access to the latest and most advanced technologies, including glass from northern Europe, and it may have been through experimentation with a lens or mirror that Brunelleschi discovered linear perspective. This suggests that the technologies we use affect the way in which we see and comprehend the world.

In the last century, a wide cross-section of society has started to view the world dynamically through a frame. We can sit and look at the cinema screen, the television, car windscreen, computer monitor or games console and watch our viewpoint move rapidly through the world. Experience in these media makes it easier for us to understand and use virtual reality. For example, car travel in the real world has similarities with our experience of virtual reality – the car restricts our perception of the world to a dynamic view through a frame. We view the world through the window and, although travelling at speed, our body remains static. The development of urban simulation may plausibly be linked to the rise of car culture and subsequent development of driving simulators. The idea of experiencing a world by simulating smooth movement through it makes sense to those who have learnt to navigate their cities sitting behind their steering wheels. It is telling that one of the first large-scale photo-realistic urban simulations was created in Los Angeles. This is a city in which buildings (such as the Chiat/Day building by Frank O. Gehry and Associates) have been created to be viewed from a car, moving past at speed (Plate 2).

Many people have gained experience of virtual reality from computer games and multi-user virtual worlds. The expectations that users have of professional VR packages is highly shaped by experience of early games, such as SimCity, DOOM, Quake and Tomb Raider; or online worlds such as those available through Blaxxun, Virtual Worlds or ActiveWorlds. The computer game SimCity has been particularly influential for built environment applications. Based on the belief that the complex dynamics of city development can be abstracted, simulated and micromanaged (Friedman, 1995), it was invented in 1987 after a games creator noticed that they had more fun building islands than blowing them up. SimCity gives players a set of rules

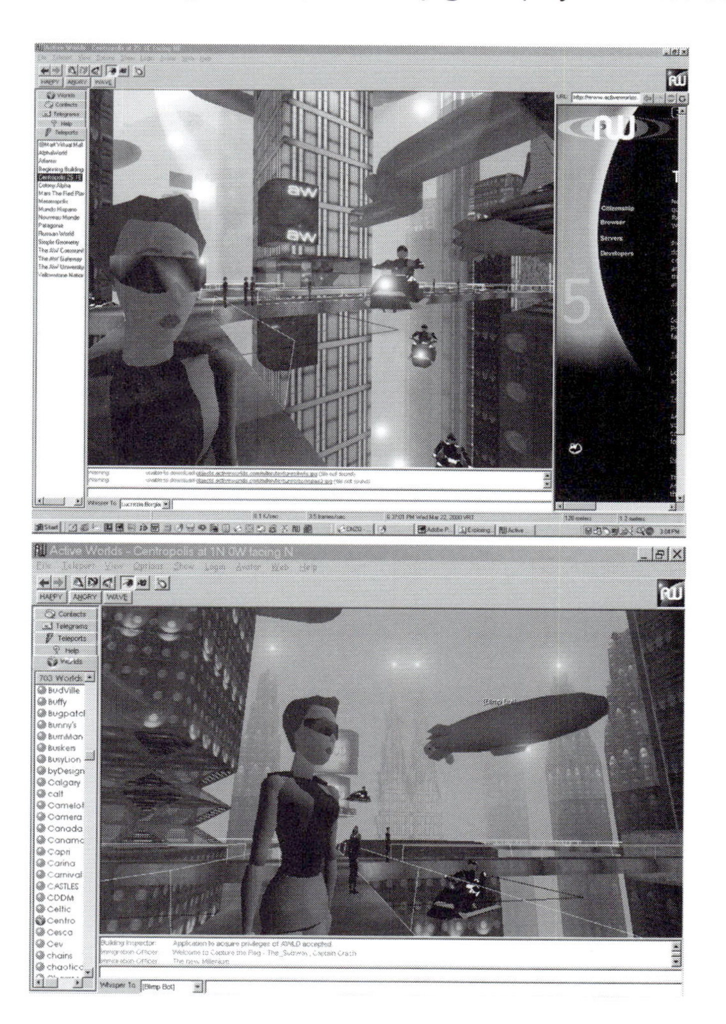

1.4
Screenshots from an online world created by ActiveWorlds

and tools that allow them to create and control a city. The player becomes the mayor and city planner in charge of city planning, resource management and strategies for dealing with disasters, unemployment, crime and pollution.

Many transferable skills have also been learnt from experience with 3D games and worlds. For example, in one of the first 3D games using a first-person viewing perspective, Wolfenstein 3D, the player moves around a building complex that is laid out on a square grid plan. Later 3D games add non-linear architecture, full use of height, cavernous spaces and models of people or 'avatars'. Some more recent games use architect-designed buildings as the games environment. The engines developed for these games are highly sophisticated and games engines are now being used to create interactive architectural models with a view to professional uses (Richens, 2000; Shiratuddin *et al.*, 2000).

By viewing the dynamic movement of the world or a representation of the world through a frame, people not only learn about using media but they also learn about the world itself. People's experiences of playing games and travelling in simulated media give them prior experience of real places. For a generation in suburban America, first knowledge of 'the city' came through television, through programmes such as Sesame Street (Pascucci, 1997). The mid-1990s can be seen as a critical period and Novak argues that:

> The technologies that would allow the distribution or transmission of space and place have been unimaginable, until now. Though we learn about much of the world from the media, especially cinema and television, what they provide is only a passive image of place, lacking the inherent freedom of action that characterizes reality, and imposing a single narrative thread upon what is normally an open field of spatial opportunity. However, now that the cinematic image has become habitable and interactive, that boundary has been crossed irrevocably. Not only have we created the conditions for virtual community within a nonlocal electronic public realm, but we are now able to exercise the most radical gesture: distributing space and place, transmitting architecture. (Novak, 1996)

For designers, the understanding of precedents is increasingly mediated through virtual reality. Mitchell points out that:

> At the very least, then, we have to admit that exploration of virtual spaces now mediates the construction of physical ones, and that physical spaces may have indefinite numbers of virtual equivalents. (1998: 208)

The availability of information and goods from other periods and geographical locations can be described as pointing to space–time compression (Harvey, 1989). An example is the experimental installation art *Northwestwind Mild Turbulence*, which was not only visited as a physical representation, but was available across the globe as a virtual representation (Plate 3). For experimental art that exists in virtual and digital form there is a question regarding which is the original – the physical installation or the virtual representation.

Issues of originality and reproduction arose in relation to art in the age of mechanical reproduction (Benjamin, 1935). To illustrate these, Berger (1972) points to the modern person seeing Leonardo da Vinci's *Mona Lisa* on a T-shirt, before (possibly) seeing the original painting in the Louvre. Such issues of originality and reproduction are relevant when we consider the built environment in the age of digital reproduction. Many people may first experience cities such as London, or Los Angeles, through computer-based car-chases along their streets on the Playstation. They bring this experience with them, and it moulds their expectations, should they visit the real city.

However, the interactive, spatial, real-time medium through which professionals, clients and end-users learn about remote architecture and urban design provides only representations and not reality. Artistic decisions have to be made about what to keep in and what to leave out of any representation. Hockney (2001) sees the period between 1930 and 1960 as an exceptional period in the history of 2D image making, as photography made the process largely mechanical and there was relatively little creative intervention. The rise of digital technologies and digital painting, etc., can be seen as a return to a greater creative manipulation of images, only now there is increased potential to create, manipulate and interact, and we are working with 3D representations.

Throughout the historical development of virtual reality, different metaphors have been used to describe its role. As shown in the next chapter, the early description of virtual reality – as though it was reality – has parallels with early descriptions of other media. We are beginning to learn more about the limitations of virtual reality as we gain wider experience of using it. Other metaphors become more relevant. We will discuss the use of virtual reality as an image and as a prototype. For organizations looking to use virtual reality, the way that different groups of staff interpret and use virtual reality affects their ability to obtain business benefit and to integrate use of virtual reality into the organization.

Development of the virtual reality system

The nature of the VR systems through which we perceive interactive, spatial, real-time representations is changing. There is a trend towards smaller, cheaper or more flexible systems, which incorporate both sensitive input devices and output devices with greater resolution. Table 1.1 shows major developments in enabling technologies, many of which have been heavily influenced by the needs of entertainment, military and advanced manufacturing applications. The heritage of VR systems is shared and enabling technologies are used in flight simulation, urban warfare simulation and CAD software, as well as virtual reality.

Pre 1950

The first real-time computer, 'Whirlwind', was developed in the middle of the last century. It was in the late 1940s and 1950s that the first digital computers such as the ENIAC (Electronic Numerical Integrator and Calculator) were created. These computers operated as giant calculators, numbers were entered and eventually an answer came back, in a process that became known as batch processing. In contrast, 'Whirlwind', which had been under development at MIT since 1944, was designed to try to respond instantly to whatever the user did at the console. Developed as part of Project SAGE – a crash programme to create a computer-based air-defence system against Soviet long-range bombers – it started out as a flight simulator and evolved into the world's first real-time computer (Waldrop, 2000). Though the Whirlwind computer had only 1024 bytes × 2 banks of memory, it was physically very large, weighing 10 tons and consuming 150 kW of power.

1950–1970

Early computers such as Whirlwind did not have sophisticated graphical interfaces. In the 1960s, a computer scientist working at MIT set out to change this and argued that 'In the past we have been writing letters to rather than conferring with our computers' (Sutherland, 1963: 8) He created a graphical system, Sketchpad, which allowed the

Table 1.1
Major developments in enabling technologies, on high-end and low-end systems

	Pre-1950	*1950–1970*	*1970–1985*	*1985–1995*	*1995–2000*	*Post-2000*
High-end systems	Whirlwind – the first real-time computer Keyboards, Cathode Ray Tube (CRT) displays Development of flight simulation	Sketchpad – the first CAD application Sketchpad III – the first 3D CAD application Lightpens, the mouse, head-mounted displays Development of computer graphics and human–computer interaction (HCI) including haptics	Walkthrough – the first interactive architectural walkthrough Advanced rendering of 3D objects (Gourand, Phong shading, etc.) Networked computing for processing complex graphics	First commercial VR SGI leads hardware market and develops 3D APIs, Open GL (Inventor and Performer) Commercial VR software from Evans & Sutherland, Multigen Paradigm, WTK, Division, etc. Peripherals such as the treadmill, BOOM and Fakespace's CAVE and Immersadesk		Motion capture Volume visualization
Low-end systems			Introduction of the mouse and joystick	Gloves, active and passive stereo, PC-based graphics cards PC-based VR software pioneered by VPL Research, Virtuality and Superscape Web-based 3D Multi-user worlds VRML 1.0 Basic translation from CAD to VR	Flat and high definition screens Mobile computing CAD to VR and GIS to VR data translation WTK and Division provide PC-based versions of their software VRML 97, Direct 3D, streamed Web technologies Force feedback sketching devices, tablets	Motion capture Volume visualization Auto-stereoscopic displays and flexible 'roll-up' screens 3D Laser scanning

drawing of vector lines on a computer screen with a light pen. This is now commonly referred to as the first CAD package, though it was not the only one developed at this time (Myers, 1998). Other examples of early CAD packages include DAC-1, the package used by General Motors in 1963, and Sketchpad III, a 3D CAD package developed in military-funded research at MIT (Johnson, 1963).

Immersive displays were being developed in work by military and civilian researchers in the 1960s, but much of this research was not published until later. Research on flight simulation, by the US Air Force (Furness, 1986) and NASA (McGreevy, 1990), contributed to understanding of the technical requirements for virtual reality (Earnshaw *et al.*, 1993), whilst later work by Sutherland (1965, 1968) developed the concept of the immersive 3D computer environment, viewed through a Head-Mounted Display (HMD).

Interfaces and peripherals for human–computer interaction were also developed. Englebart and colleagues first described the mouse (English *et al.*, 1967) and Brooks *et al.* (1990) pioneered haptic feedback using a touch-sensitive glove. Much of this research was interdisciplinary, with different specialists within computer science, engineering, psychology and ergonomics collaborating on the development of these technologies and their interface with other technologies used for design.

1970–1985

In the 1970s, computer graphics were greatly improved by research conducted at Utah, where Sutherland and his students explored the rendering of 3D objects. The process of creating a final image from a set of geometrical data, known as 'rendering', may involve hidden line removal, the addition of colours, textures, lights and shading. Researchers used networks of computers to get more processing power for complex 3D graphics. In this decade, the first interactive architectural walkthrough system was developed at the University of North Carolina (UNC) and this continued to be refined in a major research programme (Brooks, 1986, 1992). At this time Krueger (1991) was developing video projection methods described as 'artificial reality'.

It was in the 1980s that the processing power and graphic capabilities of low-end systems became sufficiently developed for their widespread use. In the 1980s, games on PCs such as the BBC Micro, Commodore 64 and Atari ST computers became popular. These games ran on very low-end systems, for example the game Elite, which attempted to show a 3D universe, ran on 8-bit machines.

1985–1995

It was not until the late 1980s that the commercialization of VR packages took off. W Industries was founded in the UK, and VPL Research Inc. began trading in the USA. The chief executive of the latter is credited with coining the term 'virtual reality' and the term was first used at this time. Interactive 3D became possible on the personal computer and applications were designed for low-end systems. For example, AutoDesk, Inc. demonstrated their PC-based VR CAD system, Cyberspace, at SIGGRAPH in 1989.

Throughout the 1990s, the games market continued to drive developments on low-end systems. The game Wolfenstein 3D was released in 1992 and ran on Intel 386 32-bit machines. New personal computers with graphic interfaces were released, including the Pentium and Pentium II, and inventions in graphics cards made low-end systems more capable of rapidly updating 3D scenes.

As CAD tools became widespread in industry, the transfer of files between different CAD packages and between CAD and other design software became more important. The International Alliance for Interoperability (IAI) was set up in 1994 to develop standards to support computer-integrated construction. It has built on the STandard for Exchange of Product model data (STEP) initiative, which saw STEP adopted by the International Organization for Standardization (ISO) as a formal standard, but not widely implemented in CAD packages.

Advances in underlying technologies not only led to developments in virtual reality, but also to developments in related applications such as GIS and CAD. Throughout the 1990s, GIS applications that facilitate the manipulation and analysis of information that is tied to a spatial location have developed and matured. Whilst early CAD tools had enabled 2D drafting, the more sophisticated CAD packages developed in the 1990s enabled 3D design. Object-oriented CAD allows the manipulation of objects, rather than lines. Using this, objects can be given behaviours and act as they would in the real world. Hence when a wall is moved, a window in that wall will move with it. Parametric modelling is another approach, which involves the use of mathematical variables or parameters to control, modify or manipulate design.

Many of the peripherals associated with virtual reality were first commercialized in the 1990s. The use of virtual reality increased both in high-end VR facilities and on the personal computer. The VR hardware supplier Fakespace introduced peripherals such as the CAVE (the name is a recursive acronym for CAVE Automatic Virtual Environment) and the Immersadesk, which enabled large-scale display of information.

1995–2000

At the same time, software protocols were being developed. Standard procedural models for 3D were based on the non-proprietary Open Graphics Library (Open GL). The VR hardware and software supplier Silicon Graphics (SGI) also introduced Open Inventor and Iris Performer, which provide further functionality allowing the programmer to concentrate on world creation.

In the mid-1990s, the Virtual Reality Modelling Language (VRML) was developed to provide virtual worlds networked via the Internet (Bell *et al.*, 1995). Based on Open Inventor, the first version was later extended to become an international standard VRML 97 (ISO/IEC 14772-1). Though VRML has been used in Web applications and CAD packages on many low-end systems, its early promise has not been fulfilled.

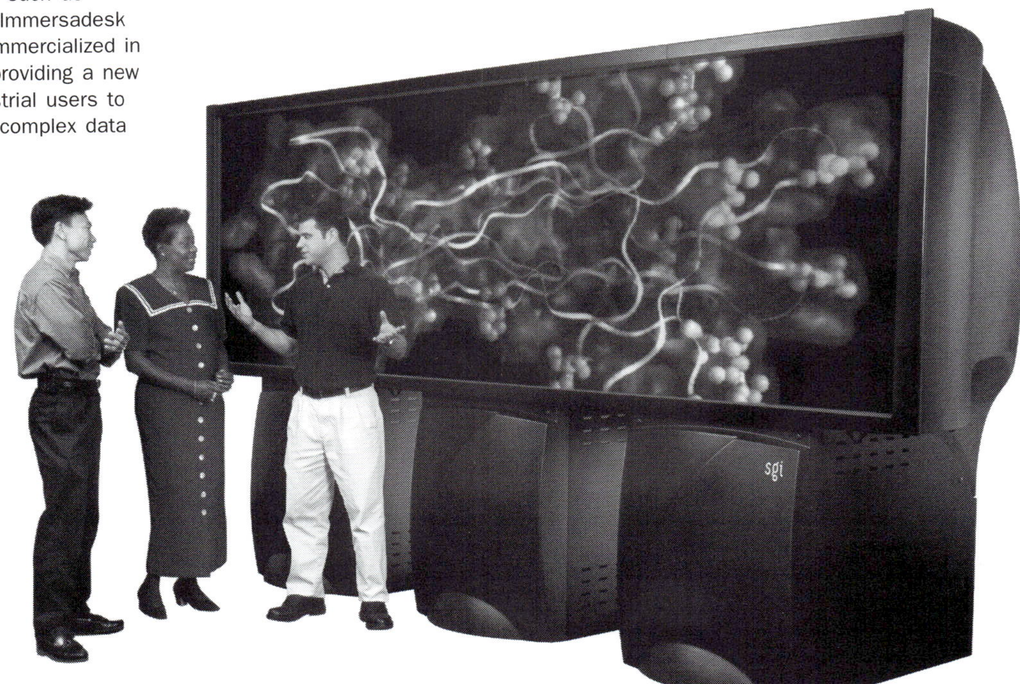

1.5
Workbenches such as Fakespace's Immersadesk were first commercialized in the 1990s, providing a new way for industrial users to interact with complex data

Whilst Silicon Graphics championed the open standards on which VRML was based, Microsoft brought out a proprietary standard for Windows, Direct3D. It was this proprietary Microsoft standard that was most widely used by PC-based games and hardware developers at the end of the 1990s.

Hopes for an open 3D graphics standard are still unrealized. As well as the competing VRML and Direct3D standards, the 1990s saw Sun Microstation's Java3D and the Microsoft's abortive Fahrenheit initiative. One problem faced by competing technologies and potential standards was the rapid rate of technological change. Even as VRML became a standard, it was being superseded by proprietary technologies, which were streamed and hence allowed users to interact without waiting for an entire model to download. At the end of the 1990s, online 3D was accessible, but it had not become ubiquitous and support for it was not standard in major browsers and operating systems. As graphics technologies continue to evolve, new standards such as the Web 3D consortium's open standard X3D have been proposed.

Post 2000

In the early 2000s, data input techniques have also been rapidly improving, particularly in the areas of 3D laser scanning and geometry capture from images and film. Models are being built from as-built data, rather than geometrical CAD data. Advances are being sought in displays, through volume visualization, auto-stereoscopic displays, and more portable and flexible roll-up screen technologies. Though built environment applications have not been major drivers for the development of these

1.6
A 3D laser scan of the Marie de Plaisir building, created using Mensi's 3D Ipsos software. The Marie de Plaisir building is the City Hall of Plaisir, in the west suburb of Paris, France

1.7
Geometry can be captured from photographs. This can be used to build the 3D model – as in this model of Canary Wharf in London, UK, which was created at University of College London (UCL)

technologies, the form that the technologies take affects them. Applications and plug-ins specifically designed for users in the construction industry are beginning to be developed. Challenges remain, for example data exchange continues to be problematic and there is a pressing need for more intuitive interfaces for working and playing with computers.

Focus on applications

Researchers have argued that virtual reality may support innovative activities by allowing experimentation in depth, involvement of all in the innovation process and an ability to capture ideas generated in the innovation process (Watts *et al.*, 1998).

Organizations are using virtual reality in conjunction with a range of other advanced technologies, such as object-oriented CAD, parametric modelling, laser scanning, photogrammetry and Geographic Information Systems

(GIS). The use of these other technologies influences the strategies that organizations develop for building and optimizing VR models and for translating data to VR systems.

Creating models

Virtual reality is just one of the possible media in which 3D data can be visualized. Ideally 3D data files should be independent of their use. Kiechle (1997) describes the basic 3D CAD data files as original artworks, like score sheets in music, that can be interpreted in different ways:

> And just like an original music score can be interpreted in a variety of ways ranging from classical instruments, jazz versions to fully synthesized versions, the 3D data file can be visualized through line drawings, computer generated still images, fly-throughs, print-outs on paper or projected images.

In practice, each application imposes its own demands on the way data-sets are constructed. For example, in virtual reality there is a trade off between the computing power available and the amount of data visualized. Models are built and optimized to make them less computationally intensive.

For professionals that are using lower-end VR systems, or that want to be able to put VR representations onto the Internet, one barrier to the use of virtual reality is the size of architectural models. One provider of models for housing developers said 'we are still working on polygon issues but hope to deploy [interactive, spatial, real-time software] in the near future'.

They point out that 3D models created for customers are large, with models of single family homes having more than half a million polygons. Large models are difficult to view on low-end VR systems unless they are optimized.

All visualizations in virtual reality are not the same. Different strategies can be used for building interactive, spatial, real-time models from 3D data and the modellers' priorities affect the visualization. The strategy chosen depends on the input data, the task and the system. The model created depends on whether it is interaction, realism or real-time viewing that is of highest priority. The system used also affects the extent to which (1) interaction, (2) realistic rendering and (3) real-time viewing are obtainable.

1. *Interaction is affected by the frame rate,* or rate at which the image of the virtual environment is being updated on the screen, and the system latency, or time required for the system to respond to user actions. The minimum requirement for interaction can be described as 10 frames per second with a latency of 0.1 seconds (Rosenblum and Cross, 1997). Basic interaction with a simple model is obtainable on mobile computing devices, whilst greater realism and interaction are available to users of high-end VR facilities.
2. *Rendering of virtual environment* can be simple, showing the polygons that make up spatial models as wire-frames, or with flat and smooth (Gourand) shading. More sophisticated and realistic forms of rendering allow lighting effects and transparency to be shown but require more computational time. On any given system, real-time virtual environments cannot be rendered to the same degree of visual realism as animations and still images, as the latter are not rendered in real-time.
3. *The potential for real-time viewing* is affected by the VR system, as well as the interaction, complexity of the model and rendering. Twenty frames per second (fps) is about the minimum rate at which a stream of still images are perceived as smooth animation (Isdale, 1998).

For built environment applications, many of the spatial models that are interacted with in real-time are created using data from CAD or GIS packages. If the data is complex and highly detailed, and sophisticated rendering is used, then the computational time required may slow user movement to an unacceptable level. Data structures in spatial models are often optimized.

Model optimization

When optimizing models that have been created, the desire for accurately detailed or realistically rendered geometrical information is balanced against the need for real-time interaction. If real-time viewing is of the utmost importance, as it is for flight simulation, then geometry can be simplified. Optimization techniques include:

- *using texture maps.* Texture maps are images that are mapped onto surfaces of objects to show the detail of their surfaces. By using texture maps the level of geometric detail can be reduced.
- *using primitive solids.* Simple objects, such as the primitive solids – spheres, cubes and cylinders – can be

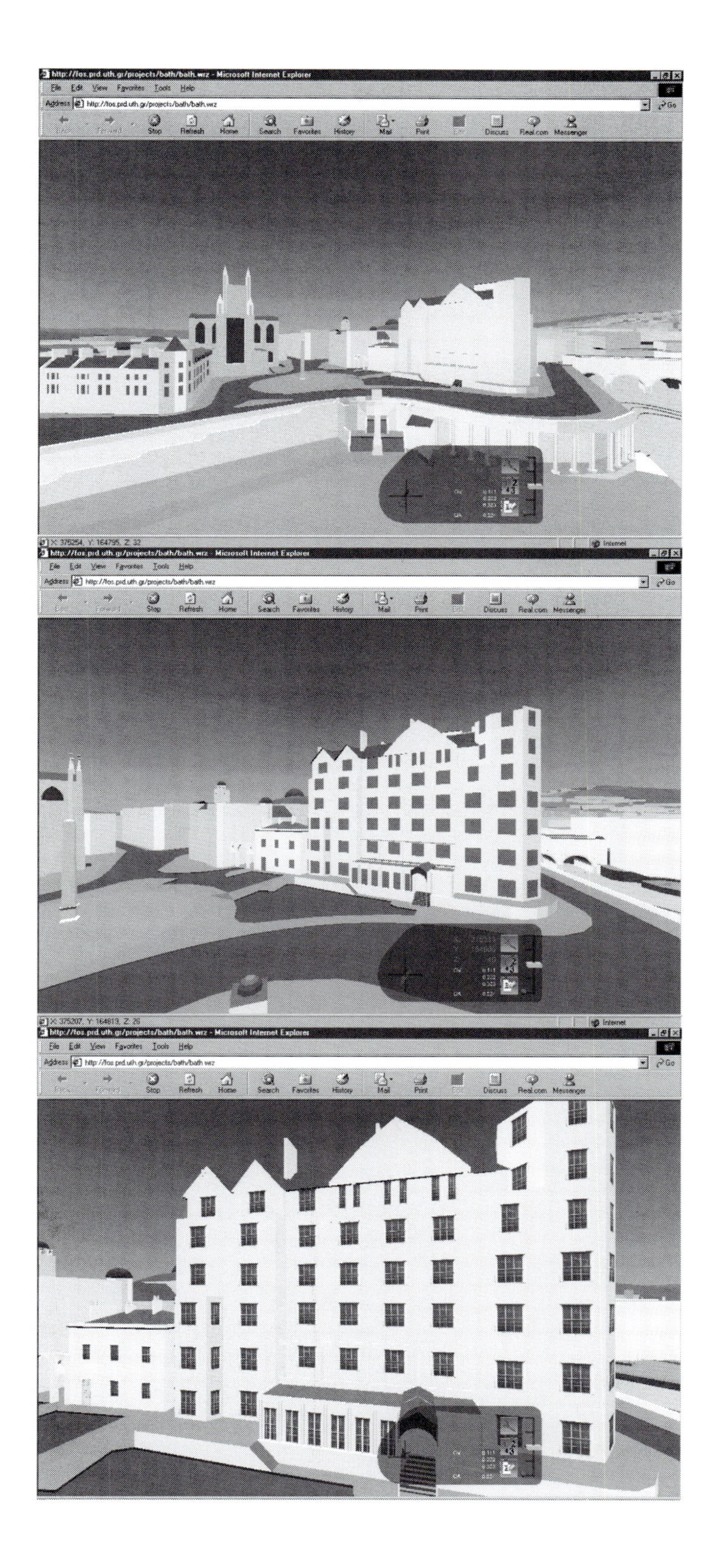

1.8
A view of the model of Bath, by CASA at the University of Bath, showing levels of detail

used together with texture maps to simplify the geometric data in a model.
- *using distance-dependent levels of detail* (LODs). Simpler geometry can be used to replace complex geometry at a sufficient distance from the viewpoint for the eye not to perceive the loss of detail.
- *using billboards*. To provide simple representations of complex objects such as trees, texture maps are used. Images of objects that are visible from all directions can be put onto billboards, which are planar objects that always face the viewpoint.
- *selectively loading objects* within the model depending on the viewpoint. Visibility sensors can be used to determine which part of the model is being viewed and therefore which geometry needs to be loaded and rendered and which behaviour scripts need to be active (Roehl *et al.*, 1997).

Optimization allows real-time viewing by reducing the information to be processed and hence reducing the computational effort required during simulation. On any system, trade-offs are made between the amount of geometric detail (number of polygons), the rendering and lighting, and the speed at which navigation and interaction are possible.

Strategies for translating data

The translation problems that plagued early users of virtual reality are diminishing. However, some thought and effort may still be required to move data from CAD into virtual reality. Strategies include building a library of optimized standard parts, relying on imperfect model conversion through translators and using virtual reality as an interface to a central database (Whyte *et al.*, 2000).

A library-based approach, where a library of components or objects is archived for reuse within the VR environment, eliminates the need for repetitive data transfer and optimization of common parts. Such an approach can also be described as object-oriented. Objects can encapsulate information about their behaviours and processes, as well as their geometry. Significant time and effort is initially required to build up the library; however this time is compensated by the reuse of information. Architects have been active in championing technologies such as Geometry Description Language (GDL), which are examples of the use of this approach.

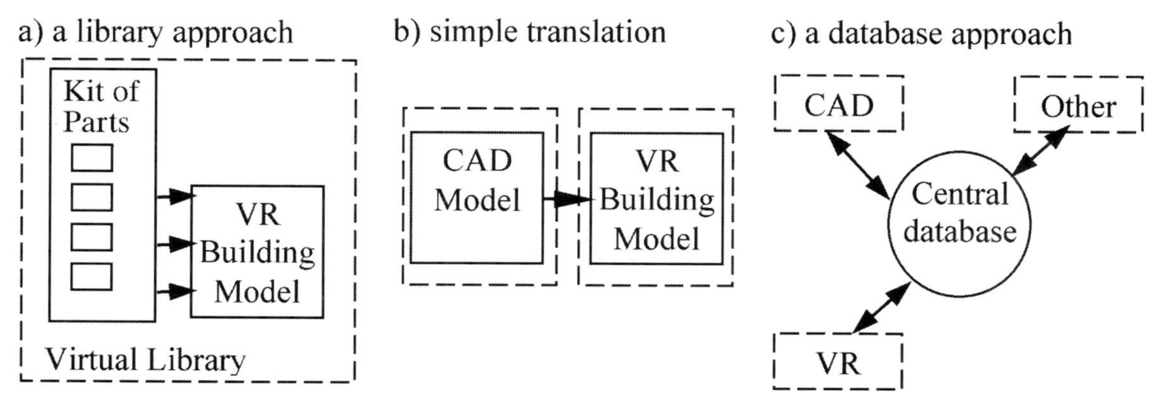

1.9
Library of forms, database and simple translation approaches

Complete CAD models can be used to generate VR models by straightforward translation of the whole model, sometimes in conjunction with algorithms for optimization. A translation approach has been used in projects where there are few repeated elements and the data is predominately geometric, or where the design process is completed and the design is fixed and unchanging. The result is typically a highly rendered or optimized model for presentation.

A database approach to VR model creation uses a central database to control component characteristics and both CAD and virtual reality are used as graphical interfaces to that database. The building model is created in the central database and viewed through the different applications, one of which is the VR package. A full implementation of such a system would allow updating of the model in both CAD and virtual reality. There are not yet any commercial implementations of this approach, but the Open Systems for CONstruction (OSCON) research project at Salford University uses case studies from real life construction projects to demonstrate its usefulness (Aouad *et al.*, 1997).

Organizations that use virtual reality for the representation of engineering and design data may or may not be involved in the creation and optimization of the models that they use. If they see the creation and optimization of models as a specialist activity, then it may be outsourced and the data visualized may be essentially offline data, which is not integrated with other digital source data. Yet, if virtual reality is to be used as a generic technology, across all the functions of an organization, then it must be used in conjunction with other advanced technologies and the data visualized must be integrated with other digital data.

This book is inspired by the leading engineering, design and construction organizations, and is based on research into their use of virtual reality and interactive 3D. We look at case studies of industrial use of virtual reality in the design, production and management of the built environment. First we will look at representations in Chapter 2 and consider how virtual reality can be useful as a representation of data in the design, production and management of the built environment.

2 Maps, models and virtual reality

In a famous story by Borges (1946), Schools of Cartography become skilful at producing large and accurate maps. Eventually they create a Map of the Empire that is so large it occupies the whole of a Province. Yet this disproportionate Map is still not realistic enough for its creators and they become dissatisfied with it. They decided to build a Map of the Empire, which is the size of the Empire and coincides with it at every point.

The Schools of Cartography are very pleased with this Map, but slowly inhabitants of the Empire realize that it has no practical use. Following generations, which are less addicted to the study of cartography, abandon the Map, leaving it to be worn down by the weather. In the deserts of the Empire there are ruins, inhabited by animals and beggars, but in the rest of the Empire no trace is left of the Geographic Disciplines.

Is virtual reality like the Map the size of the Empire or can it deliver real business benefits to organizations? In this chapter we ask why representations are useful and how virtual reality can be used to capture relevant information about the built environment and to explore possible changes to it. By understanding virtual reality within the context of representations, we can learn to use it more effectively.

Early 'pioneers' of virtual reality were in awe of their ability to build large and accurate models. They championed the idea that virtual reality was the same as reality. However, although the VR medium allows a detailed representation of the built environment to be viewed at full scale, a representation of an object is not a replica, but '... its structural equivalent in a given medium' (Arnheim, 1954: 162).

The lived-in reality that we experience every day is much richer (and messier) than virtual reality. It contains many cues that are either absent or greatly altered in virtual environments. We live in and perceive the real world through our body and its movement through space and time (Lefebvre, 1974). As we move through the built environment, we consciously and subconsciously notice visual attributes such as nodes, landmarks, paths, edges and districts (Lynch, 1960). We obtain subtle cues from

2.1
Snapshots of lived-in realities – Zhangjiakou, China

2.2
Snapshots of lived-in realities – Sheffield, UK

hearing, smell, taste and touch, as well as from temperature, humidity, breeze and social interaction.

Though we can use virtual reality to explore spaces that predate the digital era, and those conceived within digital media, we must learn to interpret what we see. Virtual reality enables us to walk through models of buildings that no longer exist. For example, ancient architecture is available online and through display screens at museums. However, learning is required to interact with such models and to understand the extent to which they are the same as the buildings that they represent.

2.3
Snapshots of lived-in realities – Montafia d'Asti, Italy

2.4
An image of ancient architecture taken from an interactive 3D model of the Theatre at Epidaurus, Greece by Theatron. The overall purpose of the Theatron Project was to apply multimedia technologies, and in particular the potential of VR modelling, to explore new possibilities for effective teaching. Focusing initially upon the history of European theatre, the project has developed a prototype multimedia module, which will allow a new and more effective means of teaching than has previously been achieved

2.5
Photograph of 'A Stage for Dionysus' interactive touch screen kiosk, Theatre Museum, London, UK

The desire to see a new medium as a replica of reality is a common phenomenon. In other media, such as photography and film, early users underestimated the differences between medium and reality. The hero in a film by Godard (1960) famously said that cinema is truth 24 times a second. Yet we now know that photography and film do not represent the real world in exactly the same way as we experience it. Like the early users of other media, early users of virtual reality had the desire to see their medium as reality. One said 'VR can make the artificial as realistic as, and even more realistic than, the real' (Negroponte, 1995: 116). Another described virtual reality as '... a magical window onto other worlds, from molecules to minds' (Rheingold, 1991: 19) A third said 'however real the physical world is ... the virtual world is exactly as real and achieves the same status, but at the same time it also has this infinity of possibility' (Lanier, quoted from Wooley, 1992: 16).

We are now learning about how virtual reality is different from reality. Though reality cannot be replicated in other media, it can be abstracted and represented. However, substantial learning is needed to understand representations in different media. Like film, animation and television, virtual reality uses a language of cuts, pans and zooms that has to be learnt, as it is not experienced in the real world.

In this chapter the use of virtual reality is contextualized through comparison with other media. We consider the function of representations, and then look at maps and models, before considering the particular characteristics of virtual reality and tools to enhance its use.

Representations

Representations of reality are abstractions. They have a special status as perceptual objects because they have been created to be meaningful (Scaife and Rogers, 1996).

How well we perform with different forms of representation depends upon our experience, abilities, strategies and motivation (Chen and Stanney, 1999). People have widely divergent strategies for understanding spatial relations and widely divergent spatial capabilities. Different types of representation may be more useful and intuitive to different people, and experience with a medium is one of the key factors that affect the extent to which people will find that medium useful. Experts, for example, appear to have greater ability to aggregate or 'chunk' information (Simon, 1979), and use more abstract forms of representation. We may find that expert users of virtual reality use the medium in different ways than novice users.

As discussed above, advocates of new media have often tried to justify them by arguing that they are the same as the real environment. However, objects and their representations are not identical. The function of a representation may be to allow us to see space anew. Structurally equivalent representations in different media allow us to see different information about a problem. Thus the choice of medium and the generation or use of a representation may be part of the process of decomposing a problem. In this section we look at the relationship between the object and the representation, examine the role of representations in problem solving, and look at landmark, route and survey knowledge.

Objects and their representation

Whilst virtual reality can be used to enhance understanding of the built environment, there may not be a one-to-one translation between an object in the built environment and its representation in virtual reality. We cannot assume that there is a one-to-one relationship between what is signified and the signifier. The same object may lead to multiple representations. Conversely the same representation may be interpreted in multiple ways. We can see the process of representation and interpretation as an act of knowledge construction (Macheachren, 1995) or as a complex form of reasoning (Bosselmann, 1999).

For the recording of a pre-existing reality this lack of absolute correspondence between reality and its representation can be seen as a limitation. Thus, any representation, short of an identical copy in the same medium, has '... by its very nature its limits, which its user must either accept, or try to transcend by other means' (Gombrich, 1982: 173).

Yet for thinking about and looking at problems, it may be the abstract and partial nature of representations that makes them useful. Differences between virtual and built environments may be used to illuminate hidden structures. In this way, we can see the unrealness of virtual reality as a feature of the medium that can be used to advantage rather than something to be fixed by future work.

Objects in the built environment can be represented in an iconic manner, so that the representation looks like the thing that it refers to. Alternatively they can be represented in a symbolic manner, using a learnt code to represent some aspect of the object represented. Whilst the term 'virtual reality' tends to conjure images of highly realistic iconic representations of buildings, practitioners and researchers are developing tools and techniques for using virtual reality in a more symbolic and abstract manner. The greater abstraction available in symbolic representation offers a high degree of ambiguity, allowing the viewer to question and interpret what is seen (Radford *et al.*, 1997). Visual abstraction offers advantages over photo-realistic rendering for some applications (Boyd Davis *et al.*, 1996). We will see that leading industrial users are using both iconic and symbolic representations.

Problem solving

Representations are among the principal tools we have for exploring, manipulating and conjuring possibilities (Tufte, 1997; Groák, 2001). Any particular representation makes certain information explicit at the expense of information that is pushed into the background and may be quite hard to recover (Marr, 1982). On a map of a city, for example, an experienced map-reader may find information about spatial configuration and routes easy to understand, whilst they may find the visual appearance of streets harder to recover.

Good graphical representations can reduce the amount of effort required to solve problems simply by acting as external memory. Our short-term or working memory holds the information we are currently using. When we need to remember a lot of things to perform a task, we may attempt to over-fill our short-term memory and then we find ourselves forgetting things that are relevant to the task (Johnson, 1998). External representations reduce the amount that needs to be remembered. One of the engineers interviewed by Schrage summed this up 'My brain was too small; I needed external versions to see what was going on' (2000: xvii).

For any task, a good representation will also make problem solving easier by reforming the problem domain, whilst maintaining its abstract structure. The representation can make explicit the problem state thus reducing the amount of cognitive effort required to solve it. A good example is the Arabic, binary and roman numeral systems (Marr, 1982). Arabic numbers make explicit the number's decomposition into powers of ten, and hence make the discovery of numbers that are themselves a power of ten easier. Numbers that are the power of a different base, such as two, are more difficult to find. Conversely, binary numbers, which are widely used in computing applications, make explicit the number's decomposition into a power of two and numbers of base ten are more difficult to find. Roman numerals have now largely fallen into disuse as most common problem-solving activities such as multiplication and division are easier when using Arabic numerals rather than roman numerals.

For activities such as design, where the problem cannot be clearly defined, we need representations that will help us move between focused reasoning and free association (McCullough, 1998). Representations constrain the inferences that can be made about the represented world, focusing attention on particular factors (Scaife and Rogers, 1996). Ambiguity in representations may aid creative thought and moving between representations may help us to see different aspects of design problems.

Our experience of travelling through cities at speed, in the car and in new media such as film, computer games and videos, has led to increased interest in time and movement as design generators. The duration, flow, pace and rhythm of our actions affect our understanding of the city (Borden, 2001). Dynamic forms of representation are becoming necessary to explore our changing understanding of space. Venturi, Scott-Brown and Izenour argue that:

> The representation techniques learned from architecture and planning impede our understanding of Las Vegas. They are static where it is dynamic, contained where it is open, two-dimensional where it is three-dimensional – how do you show the Aladdin sign meaningfully in plan, section, and elevation, or show the Golden Slipper on a land-use plan? (Venturi *et al.*, 1972: 15)

2.1 Jyväskylä Music and Arts Centre, Finland

Designers are using a range of media in the design process. For example, when designing a competition entry for the Jyväskylä Music and Arts Centre, the architect Johan Bettum worked with both digital and physical models. The design aimed at integrating the cultural and public programmes of the project with each other and the urban setting of the town. The use of particle streams was introduced as the basis for project design, and this idea was worked out in a set of virtual and physical models.

By working across different media, the designer worked to reinterpret the role of an arts institution in the age of electronic media and mass entertainment. The idea was to create a radically open and at the same time protected space for cultural exchange.

2.6
One of a set of virtual models of the Jyväskylä Music and Arts Centre. It was constructed digitally from the corresponding subset of information contained in the initial particle cloud

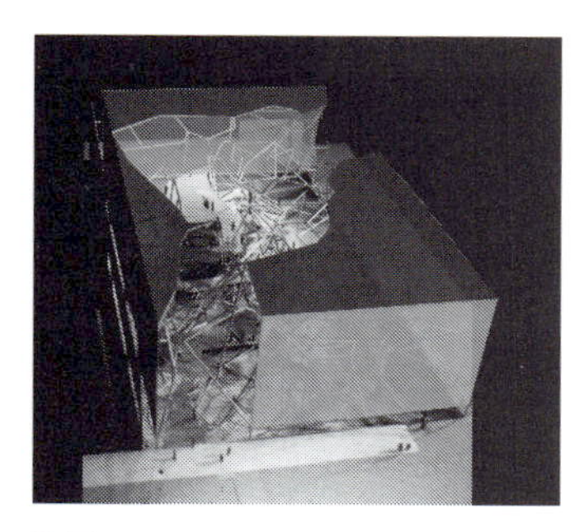

2.7
Model from the design process of the Jyväskylä Music and Arts Centre showing the proposal more or less complete and with a clear view of the tubular structure that configures the interior forms, constructions and spaces

Digital media offer designers new ways of exploring the changing concepts of space and time. Complex spatial forms are being developed and these demand new forms of representation in order to be understood. Asymptote Architecture began to use virtual reality when working on the Los Angeles West Coast Gateway project 'Steel Cloud' in 1989. Hani Rashid, of Asymptote, described his early interest in virtual reality:

> I thought to myself if I can construct this in VR, I can then convey what it is about truthfully. There were all kinds of misconceptions about the project through typical modes of representation and nobody really understood the project in its entirety, it was far too complex to be simply modelled and drawn in a conventional manner. It didn't really lend itself to just typical model making, typical representation, so I started making VR as a representational device.

These practitioners have been looking for dynamic and spatial media in which to explore complex forms. More flexible forms of representation are made available to designers through digital media (Mitchell and McCullough, 1995). Novak argues that 'Learning from software supersedes learning from Las Vegas, the Bauhaus, or Vitruvius' (1996).

Landmark, route and survey knowledge

The way that adults use representations to learn about the built environment is not well understood. The psychologists Piaget and Inhelder (1956) inspired general theories of spatial knowledge acquisition, though they were specifically interested in how a child first learns about space. They saw understanding being built up hierarchically, through a sequence of stages at which different elements of spatial knowledge are learnt. According to Siegel and White (1975), the learner first acquires 'landmark knowledge', recognizing only the patterns and characteristics that identify specific key places. At this stage they can identify key locations, but cannot navigate with confidence between them. Then they acquire 'route knowledge', which is also known as procedural knowledge. At this stage they have a familiarity with routes, knowing procedures for navigating between known landmarks within the environment, but they do not know how these separate routes are related, and they cannot find shortcuts between arbitrary points along the routes. Finally they attain the highest form of spatial knowledge, 'survey knowledge', which is also known as configurational knowledge. This is when they know the environment fully to the extent that they can draw a map and understand the direction and distance of any location.

Theories based on this developmental approach have been widely applied to describe adult learning of space (e.g., Siegel and White, 1975; Wickens and Baker, 1995). However, there is little evidence that the hierarchical sequence proposed accounts for how an adult learns about unfamiliar places. Adults already possess developed spatial skills and even if the learning sequence landmark–route–survey is found in children, it may not be found in adult spatial learning. Theories based on this approach also fail to adequately explain the use of abstract representations. Indeed, if we consider an adult studying a map to learn about a new city, the hierarchical sequence appears to be quite false. An adult may attain survey knowledge without acquiring either knowledge of landmarks or familiarity with routes. It is questionable whether this adult, who may never even have visited the city itself, can be described as having thus attained the fullest form of spatial knowledge.

The three elements, landmark, route and survey knowledge, provide a useful way of breaking down spatial knowledge into different components that can be separately considered. Representations vary in the extent to which they make explicit these elements of spatial knowledge. Though these types of knowledge are not built up in a hierarchical sequence, we can understand adult learning of space as comprising an increase in extent, accuracy and completeness of each element.

Maps and models

Many representations are used in the production of the built environment and these span a wide range of 2D, 2½D and 3D media. They include paper-based plans, sections and perspectives, cardboard models, computer-based simulations and animations, as well as interactive, spatial, real-time models.

Different types of representations make explicit different sets of information about the real world, and we can expect these representations to have various strengths and weaknesses when used for different tasks. For example, we may find that using a map to evaluate the best route between two points is easier than using it to imagine a street scene. Scale may also be an important aspect, and the structures we see at different scales may differ.

To contextualize our discussion of virtual reality, we will first look at the use and characteristics of 2D, 2½D and 3D forms of representation. We will then consider virtual reality.

Two dimensions, 2D

Two-dimensional (2D) representations, such as maps and plans, allow us to see an environment that is too large and complex to be seen directly (Macheachren, 1995). Three-dimensional phenomena are simplified through abstraction into 2D. Henderson writes that 'To see a map is not to look from some imagined window but to see the world in a descriptive format' (1999: 28).

The representation can be used as a quick way of gaining knowledge about the configuration of an existing environment. It is used as a kind of shorthand in design. In 2D representations, a whole environment can be simultaneously understood from a single vantage point. The ability to look at the world at different scales, such as 1:500, 1:200, 1:100, 1:50, 1:20 and 1:10, allows structures that are most apparent at these different scales to be considered, and moving between scales shifts the focus of attention.

2.8
Maps have been used for many years to record and explore urban form. This 1931 Ordnance Survey map of Southampton shows the layout of the city, with contours (at 10 feet vertical intervals) and scale shown in miles and feet

The architects von Gerkan, Marg and Partners have described how the easy variation of scale provided by the photocopy zoom facility was useful to them in conceptual design. They found that the precision of the 1:200 scale, which is used in competitions, could be a hindrance in the design of large projects because insignificant details were considered and drawn at an early stage before the fundamentals were solved. Using the photocopier, they could design at 1:400 and enlarge the drawings afterwards for presentation, thus also gaining drawings that had the advantage of graphic succinctness (von Gerkan, 2000).

The way we understand space in 2D spatial representations is not the same as in the physical world. The medium of acquisition of spatial knowledge affects the spatial knowledge achieved. In a famous experiment, employees with *in situ* experience of a building were significantly more accurate at estimating directions and route distances than a group of participants who had only studied its floor plan (Thorndyke and Hayes-Roth, 1982). However, the map participants made straight-line distance estimates that were more accurate than new employees and as accurate as those with 12–24 months' experience of the building. Maps and plans provide a means of rapidly assimilating knowledge about the relationship between different parts of a building, making straight-line distances explicit, but providing less explicit information about direction and route distance.

Two and a half dimensions, 2½D

Representations that show three spatial dimensions projected onto a 2D plane have been described as 2½D (Marr, 1982). These include perspectival representations and those using parallel projection systems, such as axonometric or isometric projection. Whilst parallel projections are described relative to a frame of reference based on the principal axes of the object itself, in perspective the objects or scenes are drawn from a particular point of view (Dubery and Willats, 1972). Henderson argues that perspectival illustrations play only a supporting role and do not contribute the optical consistency that is crucial to creating an object from a drawing. In a case study an engineer replied 'Would you want a dress pattern in perspective?' (1999: 33).

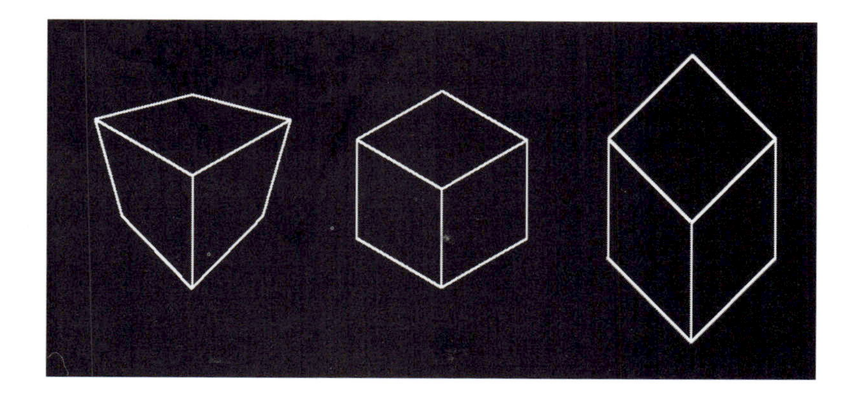

2.9
Representation in perspective, axonometric and isonometric

Three dimensions, 3D

Three-dimensional representations or models allow us to see spatial aspects of the existing and proposed built environment. Since the Renaissance, physical models of new buildings have often been seen as a necessary accompaniment to drawings. The Renaissance architect Brunelleschi presented a model of the chapel at Santa Croce in Florence to the project patrons, the Pazzi family (Vasari, 1568). By the seventeenth century, the architect Wotton (1624) argued that no one should build on the basis of a paper drawing such as a plan or a perspective, but rather they should see a model of the whole structure at as large a scale as possible. In more recent times, the use of 3D models to support collaborative design work, rather than simply presentation of the final design, has been prompted by a desire to increase public participation (Lawrence, 1987).

Models are often categorized as physical models and computer models. However, because of the characteristics of screen technologies, computer models viewed through single or dual 2D screen systems are essentially 2½D representations. Though they are created in 3D, they are viewed in 2½D. They are seen as perspectival, axonometric or isometric representations on a single 2D viewing plane or dual 2D viewing planes. Thus, in media such as virtual reality, the description of representations as models is contentious. Though virtual reality is described as an interactive, spatial, real-time medium, it is viewed as an inherently 2½D representation.

2.10
Views of the physical model of Asymptote's Los Angeles West Coast Gateway project 'Steel Cloud'

We can also categorize models as static, dynamic or interactive. Static models include physical models and the 3D models in traditional CAD packages, which can be viewed from any number of essentially static viewpoints (though these viewpoints can be scaled dynamically). Dynamic models change over time and are viewed through film and computer animation. Interactive models are computer generated; those that are viewed in real-time are described as virtual reality. Before considering interactive, spatial, real-time representations, in the next section we will consider static and dynamic 3D models below.

Like plans, static physical models can be built at different scales. Perceiving the whole 3D environment under consideration from one viewpoint may offer a cognitive advantage. The ability to appreciate the link between a scale model and space to which it refers seems to be quite basic to humans. Research on children's understanding of simple scale models suggests that young children develop an understanding of the basic symbolic correspondence between models, maps and their referent spaces at around the age of three years (Freundschuh, 2000).

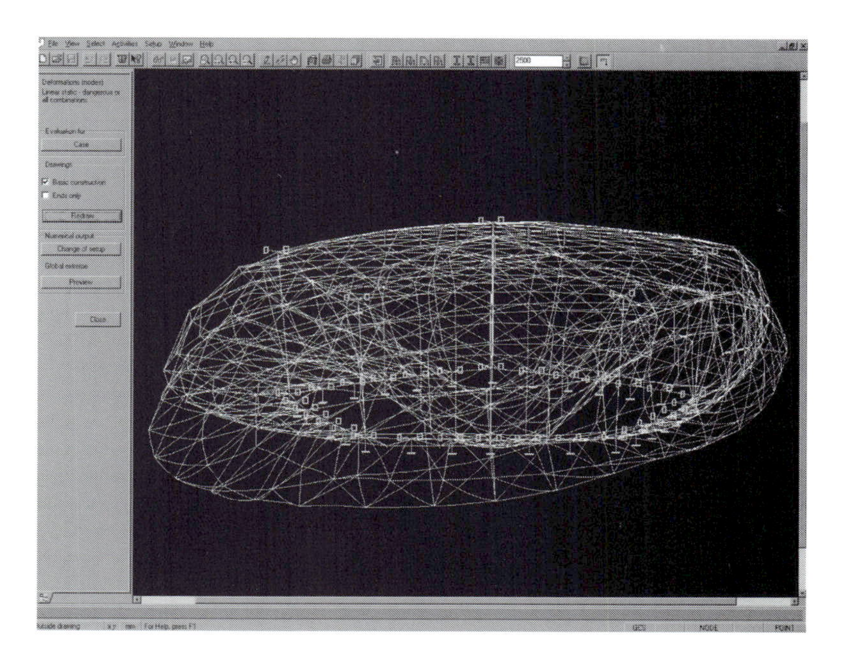

2.11
Three-dimensional design being conducted using the computer. This computer model is part of a project called 'trans-ports' by Oosterhuis

Design and visualization managers in construction companies argue that scale is irrelevant in digital models. The scale rule is redundant as design can now be conducted at full scale, a scale of 1:1. However, the spatial dimensions used in modelling are different from those used in viewing the model. It can easily be viewed in its entirety on the computer screen, or printout. If the image is too small it is an easy operation to 'zoom in' and does not involve any alteration of the scale the drawing is being drawn at. The scale at which the model is viewed on the computer screen is usually not explicit and hence is unknown to the user.

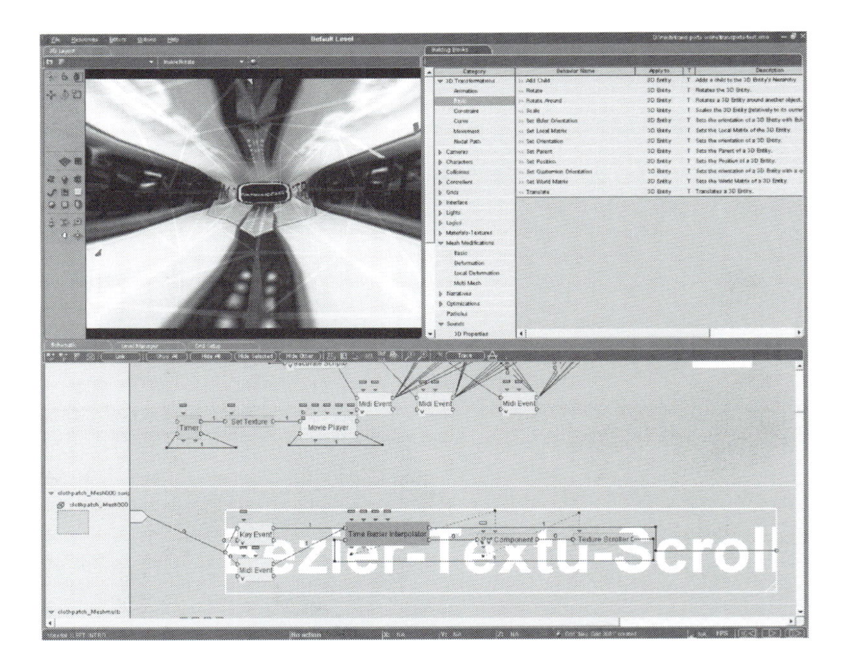

2.12
Dynamic 3D models are now widely used by architects and designers. This figure shows the Oosterhuis project 'trans-ports' being animated in the computer

Dynamic 3D media include film and computer animations. The fact that people are capable of assimilating spatial knowledge through this type of simulated media has been demonstrated through the work of the psychologists Goldin and Thorndyke (1982). People who had been on a bus tour of an area were compared with those that had experienced only a filmed auto trip of the same area. Though people gained spatial experience from both the bus tour and the film, they had different knowledge of the area. Participants in the film group identified tour locations and could remember the sequence of locations better than those who were on the actual tour. However, they performed less well in the orientation test. Though there were some participants in both groups that were completely disoriented, there were significantly more subjects in the film group who were disoriented than in the real tour group. No differences were found for estimation of the distance along the route or of the straight line distance between points.

Understanding virtual reality

Virtual reality is a simulated medium in which we can interact with a virtual model in real-time. In Chapter 1, it was described as having a minimum of three spatial dimensions but, as a viewing medium, virtual reality is inherently 2½D. In the 1960s, a pioneer of VR technologies, Sutherland, described his aim as to '... present the user with a perspective image which changes as he [*sic*] moves' (1968). Thus, real-time interaction with a 3D model is achieved by updating static perspectival images at a finite rate (Edgar and Bex, 1995). Interaction with virtual reality can be through different types of viewing perspectives and navigation modes.

In this section we look at the viewing perspectives and navigation modes in virtual reality, the differences between virtual and real space, descriptions of virtual reality as image and prototype, performance aids and the use of virtual reality as one medium among many.

Viewing perspectives and navigation modes

More than one type of representation is obtainable in virtual reality; we can consider different viewing perspectives and different navigation modes.

Within both immersive and non-immersive VR systems, three different types of viewing perspectives can be described:

1 *viewer-centred* (egocentric) – the user experiences the virtual world from a perspectival viewpoint similar to that through which we view the real world. If they use an avatar, which is a virtual representation of the self, then other people can see their position;
2 *centred on another object within the model* (exocentric) – The viewpoint can be disembodied and directed at an object within the world, this is most often a mobile object such as the user's avatar (Plate 4);
3 *outside the model and centred on the model itself* (exocentric) – the user can become an external observer manipulating the world in front of a static viewpoint.

There are limitations to the viewpoints obtainable in the real world. In merged virtual/real systems the viewpoint within the virtual model is combined with real images. This means that it is usually only possible to gain viewer-centred perspectives on the virtual model.

2.13
A viewer-centred (egocentric) viewing perspective – moving through Hypo Vereins Bank by Perilith

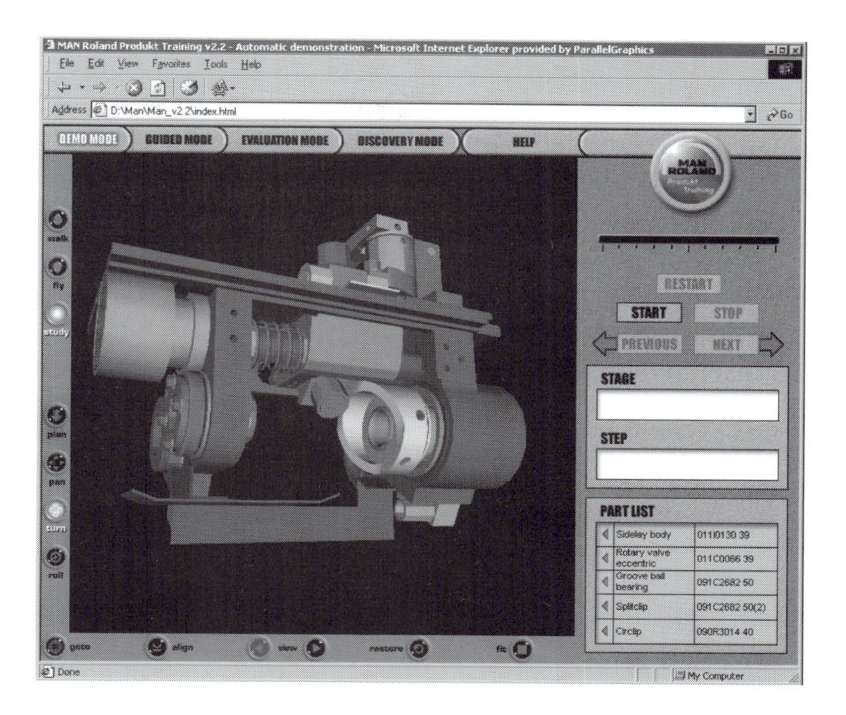

2.14
A viewing perspective outside the model and centred on the model itself – a mechanical engineering application by Parallel Graphics

Users can interact with models in virtual reality by walking through the model, turning the camera around the viewpoint, zooming, panning, orbiting the camera around the focal point, rotating the modal around the focal point, flying the camera through the scene or viewing the model on a turntable. The most common mode of interaction is flying through the model, but other types of interaction may improve performance on some tasks.

Differences between virtual and real space

Virtual reality cannot be naïvely conceived of as reality, as there are many ways in which virtual reality masks or distorts underlying realities. The characteristics that distinguish experience of virtual reality from experience of reality are both intended and unintended. They can be divided into implementation errors, limitations of the current technology and intrinsic qualities (Drascic and Milgram, 1996) though categorization of particular errors may be contentious. These distortions are similar to looking at the world through different lenses. People can adapt rather quickly to miscalibrated systems and the problems raised are more acute for mixed reality, where virtual images are overlaid on the real world (Drascic and Milgram, 1996).

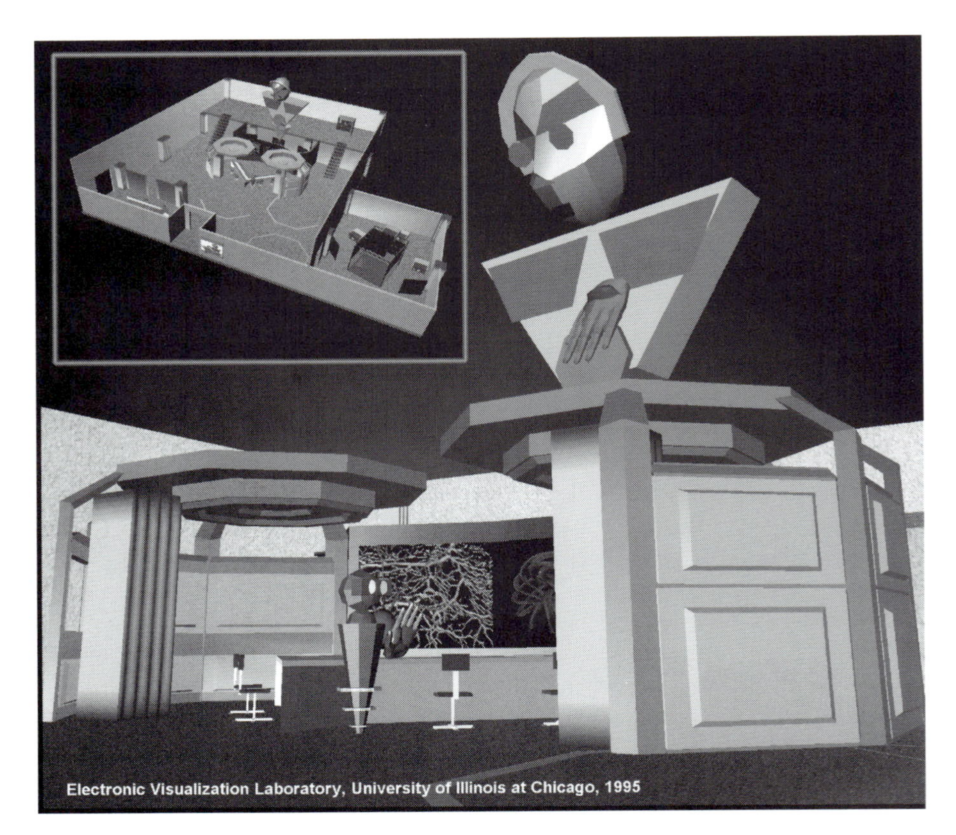

2.15
Images of 'deities' and 'mortals' in the CALVIN software

2.2 CALVIN at the University of Illinois, USA

Use of different viewing perspectives was explored in a research project at the University of Illinois in Chicago (Leigh and Johnson, 1996). The project, called CALVIN (Collaborative Architectural Layout Via Immersive Navigation), introduced two different perspectives: these are the mortal view (viewer-centred) and the deity view (outside the model and centred on the model itself). Mortals were totally immersed within the environment in a CAVE, which is a room-sized cube with stereo images projected on the walls and floor.

Deities looked down on an aerial view of the world, presented on a horizontal viewing surface on which a stereo image is visible, called an Immersive Workbench. Whilst mortals were capable of performing fine manipulations, deities were more capable of performing gross manipulations or structural changes to the world. Though the intention was that mortals and deities could assume the roles of apprentices and teachers or clients and demonstrators, the rigid use of different viewpoints was found to inhibit shared understanding of the design (Leigh and Johnson, 1996).

Though virtual reality has been described as an interactive, spatial, real-time medium:

- *interaction is not the same as action*. Embodiment in virtual reality is problematic and at best partial. The movement and actions of the body are constrained and distorted. Our experience is disembodied, as our bodies do not move but the world moves in relation to our body;
- *virtual space is not the same as place*. Though models may be created at a scale of 1:1, the scale at which they are presented is often not the scale at which the buildings and infrastructure that they represent exist in the real world. Virtual reality is not fully spatial; it is a 2½D medium and subject to optical distortions. Three spatial dimensions are represented and some practitioners and theorists argue that such models may be 4D, with three spatial dimensions and time, or even multi-dimensional. However, virtual reality is usually viewed as a projection onto a 2D screen and can thus also be considered as a non-spatial medium;
- *real-time is not the same as time*. The pace at which we experience virtual worlds is quite different to the pace at which we experience the real world. We can hyperlink to new views, fly, jump, zoom and rotate the virtual world in front of us. Though virtual reality is described as 'real-time', this term refers only to processing of interactions, not necessarily to the time it takes the user to perform comparable tasks in virtual and real worlds. The time taken to perform actions, such as motion between two parts of a city, and the physical movement required to carry out these actions are quite dissimilar to the time and movement required in the real world. For example, 'walking' between two points in a virtual model may be much faster than would be possible in the real built environment.

Our understanding of the built environment is distorted in virtual reality. Yet virtual reality is useful for representing the built environment and considering potential changes to it precisely because it is not the same as reality.

Knowledge acquired from navigating through VR models appears to be similar to, but less accurate than, the knowledge acquired from navigating through the real world (Witmer and Kline, 1998). It is different from survey knowledge acquired from exocentric views such as maps or figures (Turner and Turner, 1997). Users' ability to judge directions and relative distances from an egocentric viewpoint in a virtual building model is similar to that in the real building, improving with increased exploration (Ruddle *et al.*, 1997). However, judgement of absolute distances in virtual environments is inaccurate (Henry, 1992). Underestimates have been attributed to the blinkered nature of the field of view, which is typically 60–100° in desktop virtual environments (Ruddle *et al.*, 1997). Thus navigation through a virtual model compares unfavourably with the use of a map for learning the navigation of a complex architectural space (Goerger *et al.*, 1998) and the use of a single egocentric view may also be inappropriate for the actual design of that space (Leigh and Johnson, 1996).

Reality, image and prototype

Virtual reality can be considered as reality, as image or as a prototype. We have seen that claims that virtual reality is the same as reality are not sustainable. Yet there are theorists that consider virtual reality as an alternative reality. In the Borges story the Map the size of the Empire falls into disuse as it has no function. Yet, Baudrillard (1983) has recast this fable arguing that, had it been written today, people would live in the map and that it would be the real world and not the map that was left to ruin in the deserts.

Some architectural theorists and practitioners are looking at virtual reality as such an inhabitable alternative reality. They describe objects in interactive, spatial, real-time media as though they existed in a new form of space, rather than in spatial representations and look at what Novak (1996) terms the 'vitality of architecture after territory'. Novak argues that:

> Cyberspace as a whole, and networked virtual environments in particular, allow us to not only theorize about potential architectures informed by the best of current thought, but to actually construct such spaces for human inhabitation in a completely new kind of public realm (1996).

By describing representations in virtual reality as though they were inhabitable space, enthusiasts omit the gap between signifier and signified, viewer and viewed, real and representation (Dyson, 1998). It is the differences between representations in virtual reality and the objects that they represent that make them useful to professionals involved in the design, production and management of the built environment. As an alternative reality, virtual reality fails to be useful in the critique of real spaces.

Conceiving of virtual reality as image likewise isolates an engagement with representations from questions about the built environment that it is used to represent. It focuses attention on the consumption and production of images. Robins argues that:

> Generally we may see image technologies as still being 'in touch' with reality. But they may also be mobilized as intoxicating and narcotic distractions or defences against the vicissitudes of reality. And at their most extreme, they may be used to construct alternative and compensatory realities (1996: 123).

It is as a prototype, with a known relationship to an existing or potential built reality, that virtual reality is useful for the negotiation of contended or conflicting potential realities. Many leading users stressed the relationships between visualization and data. Virtual reality is a process tool rather than a finished product itself. It is links between the representation in VR and the reality that make it useful for design visualization within the project-based firm. Schrage (2000) points out that when organizations cannot cost-effectively couple models to reality they waste time, effort and money.

Performance aids

There are many different types of possible interaction with models in virtual reality. As we have seen, there are three different types of viewing perspectives and there are many different modes of navigation. In addition, VR packages give the user a range of tools to enhance performance, and these include access to maps, exocentric views, markers and system-wide indicators. The user may have the ability to select parts of the model, and to choose to display the whole model or only those selected parts. They may be provided with a choice of perspective or orthographic cameras, and a separate window with a plan view that can be toggled off and on. Some packages also provide the ability to measure details and dynamically take section planes through the model, generating sections at any location through the model.

Such additional information may be helpful to users of simulated media. In the experiment by Goldin and Thorndyke (1982) additional information was given to some of the participants in the bus tour group and some of the participants in the film group. Some participants additionally received a verbal description during the tour, others were allowed prior map study whilst a third sub-group on both the bus tour and in the film group received no additional information.

The results were surprising. They show that adding an additional navigational aid to the film has task-dependent consequences. In one task we see that having a map or narration hinders performance, while in another task the map improves performance and narration lowers it. So not only is the medium that is most effective task-dependent, but the type of aids that are useful in simulated media are also task-dependent. In the experiment described, the narrative provided during the tours gave names of the streets on the route, landmarks, the distance between intersections and the current compass direction. The map study consisted of looking at a map with landmarks and routes for ten minutes prior to taking the tour.

The findings of experiments such as that described above suggest that it is unlikely that a task-independent or user-independent set of optimal navigation aids will be found for the use of VR models.

2.16
Navigation aids – the interface to a VR application may include navigation aids. This application, produced by Parallel Graphics, enables users to explore the design of their house before purchase

The additional information available in a virtual world (as opposed to a map) increases the difficulty of performing some tasks. In one experiment, participants in the virtual environment group failed to filter out non-essential information and were quickly saturated with facts, many of which were superfluous to the task (Goerger *et al.*, 1998).

Novice users often have difficulty navigating complex virtual environments. Differences between virtual and real environments affect their performance in simple tasks such as navigation and way-finding (Satalich, 1995). They may veer off course, become disoriented or bump into virtual objects. Because of the difficulty they face in maintaining knowledge of their location and orientation they may devote considerable effort to this rather than the task-specific objectives (Darken and Sibert, 1993). The medium is new and there are few established aids or established users. Early graphical user-interfaces to immersive VR applications often included a virtual hand, which the user could see represented in the virtual space; however, manipulation of this hand was not standardized and the functionality, such as point-to-fly or point-to-select, was different

2.3 Helsinki city map, Finland

Virtual reality is being used to enhance navigation in the real built environment. The interface to 3D games is commonly understood, and there is commercial interest in using a similar interface to help navigation in the real city. Suppliers argue that 3D visual images help people to understand route instructions more easily than 2D maps, and that 3D animation is a better communication medium for small-sized phone and mobile display devices. Wired and wireless products are being developed for tourists and business travellers. Virtual reality applications may be accessible in hotels, restaurants etc. through an Internet-based application, or through the user's mobile phone or handheld personal digital assistant (PDA) (Plates 5 and 6).

A Finnish company, Arcus Software, has developed 3D route instruction products for mobile and Internet uses, which it has showcased in Helsinki. Companies and business can use these products to help their clients find their headquarters, offices and nearby hotels. In a 3D map, the user sees the landscape and nearby buildings from a pedestrian perspective. In-route instruction products, an arrow image or a dotted line, can be added in order to help guide the user in the right direction. When using a 3D map on a mobile device, the user is able to download additional pictures when moving forward until the destination is reached.

according to applications. The design of later graphical user-interfaces for VR applications has been influenced by 2D applications and familiar elements such as the menu are used (Sherman and Craig, 1995). However, most people are not expert users of virtual reality. The function of aids may not be apparent. An example of interaction failure is the user pointing at a door to go there and instead blowing it up (Sherman and Craig, 1995).

The extent to which different aids are useful is also dependent upon the individual's abilities with these aids. Those with more advanced verbal abilities may exhibit preferences for narration as a navigation aid whilst those with more sophisticated map-reading abilities may exhibit preferences for maps, etc. (Chen and Stanney, 1999). Virtual environments can be constructed to aid navigation with the use of memorable landmarks (Ruddle *et al.*, 1997). User-defined bookmarks have also been suggested as a navigational tool (Plate 6; Edwards and Hand, 1997).

For professional tasks, tools for marking changes and detecting clashes between different components are beginning to be introduced into VR packages. More sophisticated interfaces, which allow the user to undo and redo commands and an interface that allows access to detailed information about files may help construction sector users. The ability to add comments to the model and leave an audit trail of comments or to find all elements of a particular type may enhance performance of particular tasks.

Virtual reality as one medium among many

The use of multiple representations of the same phenomenon allows a problem to be seen in different ways, and helps the user to understand how different representations relate to one another. They also allow groups of people to interact with ideas using the medium they find most appropriate and easy to use. Virtual reality can be used as one medium among many.

New media like virtual reality may be particularly useful for sophisticated and confident users such as those who have experience with similar media such as computer games. However, when used in isolation, with an unintuitive interface, they may alienate some of the less computerized sectors of the community; social conditioning may affect the extent to which people feel comfortable with their use. Some British house-builders articulate these issues when

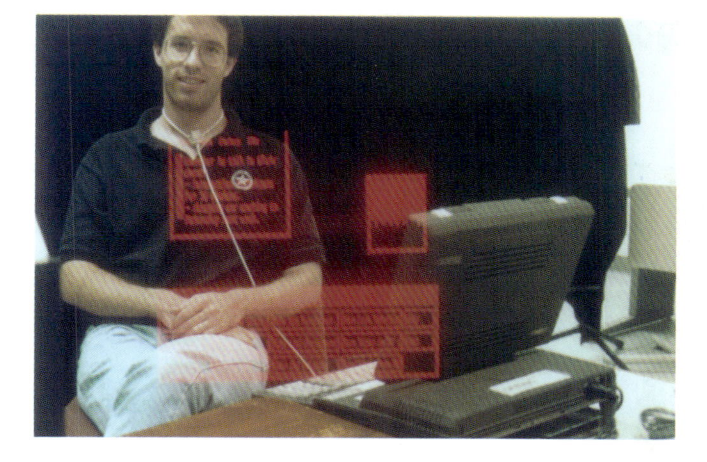

Plate 1
An augmented reality system – the image is from Columbia University's Computer Graphics and User Interfaces Laboratory, which has pioneered use of augmented reality. The user's body is tracked and applications have been created to give 2D or 3D information about their surrounding world through a transparent head-mounted display

Plate 2
Downtown Los Angeles created by the urban simulation team at the University of California in Los Angeles (UCLA)

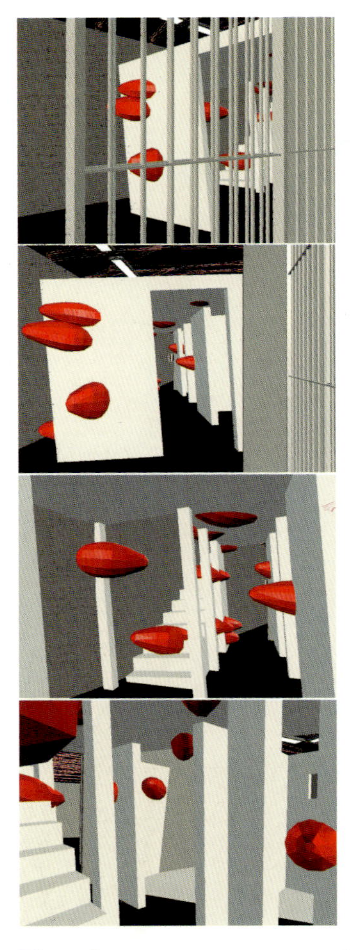

Plate 3
Views from inside the online VR model of *Northwestwind Mild Turbulence* by the artist Kiechle

Plate 4
A viewing perspective centred on an object – the viewing perspective can be attached to an avatar and move through the virtual environment with it. This virtual environment is of a peacekeeping scenario involving soldiers, civilians and vehicles. Human characters are generated by PeopleShop software, and feature expressive faces, gaze control, emotional expression and lip-synched speech. Combined with artificial intelligence provided by the Institute of Creative Technology (ICT), the characters in this Mission Rehearsal Exercise (funded by the US Army) create an emotionally authentic training environment designed to teach decision-making skills in tense situations

Plate 5
The model of Helsinki, Finland used with 3D route instruction products by Arcus Software

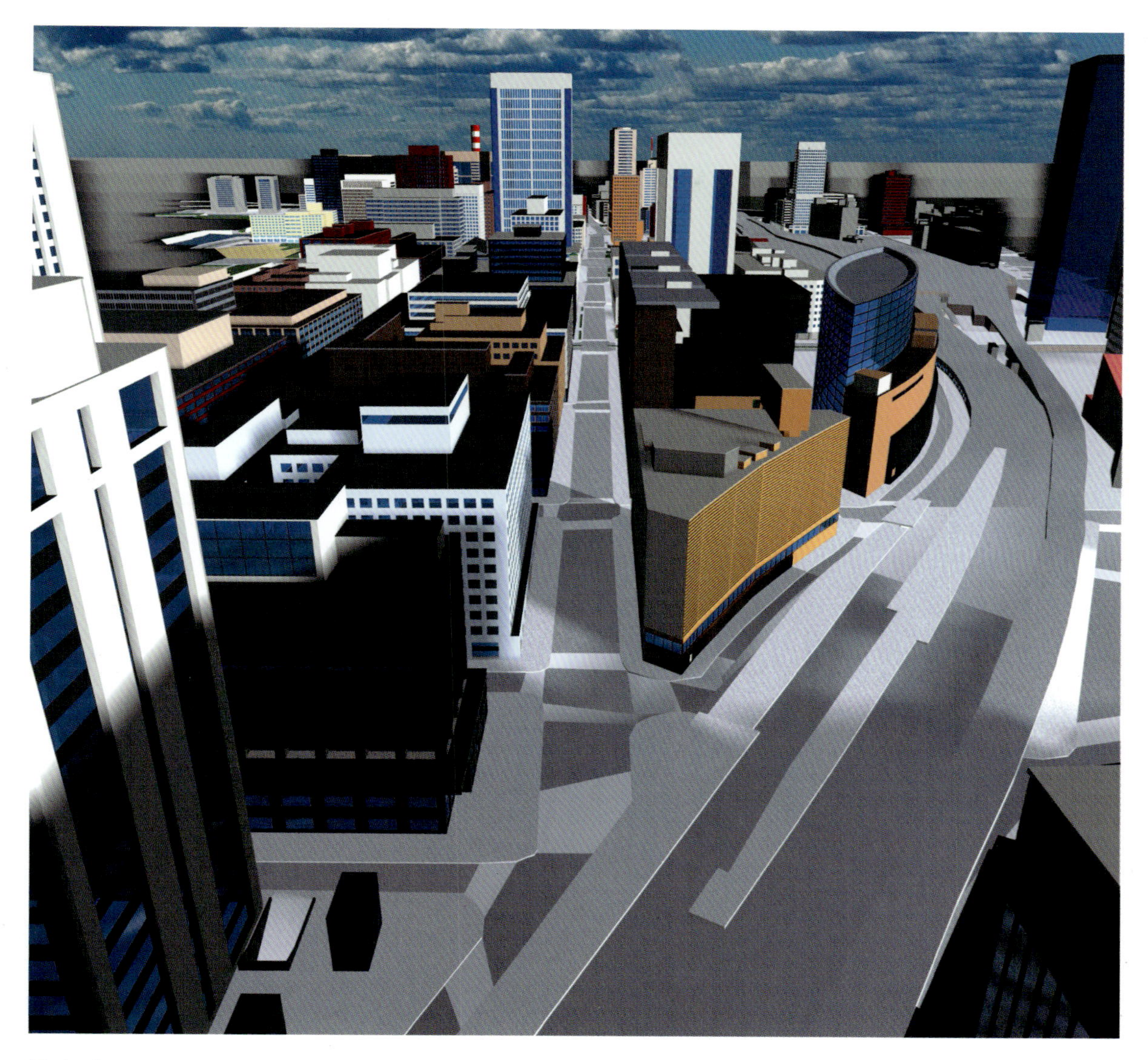

Plate 6
The model of Tokyo, Japan, which was also created by Arcus Software for 3D route instruction

Plate 7
The STEPS software, created by Mott Macdonald, in use for simulating people flow

Plate 8
Virtual reality being used for highway design on the Los Angeles 710 Freeway in California

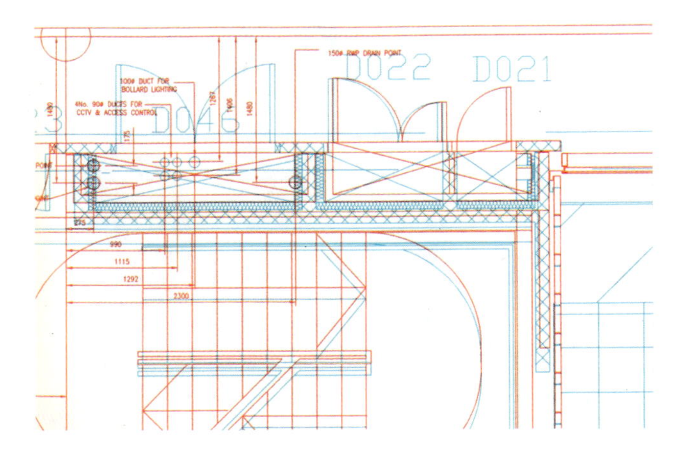

Plate 9
Superimposed design drawings showing the walls in slightly different places in heating, ventilation and air conditioning (HVAC) drawings and structural steel engineering drawings

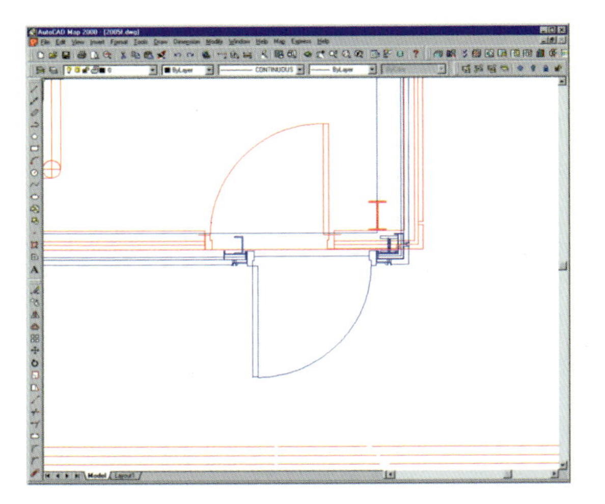

Plate 10
A door detail from the same set of drawings

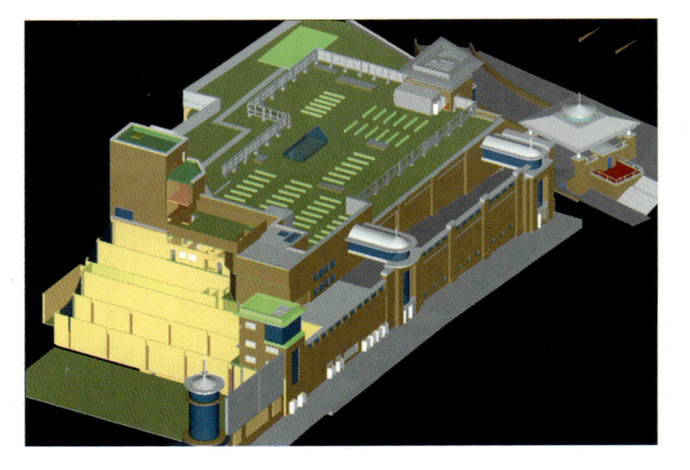

Plate 11
The full rendered model of Basingstoke Festival Place, created by Laing Construction in the NavisWorks software package

voicing their concerns about the use of virtual reality at the customer sales interface (Whyte, 2000). They perceive that key decision-makers in housing purchases may be those less interested in computer games. They want to engage the attention of all those within a group involved in a purchase and are thus careful not to introduce individual technologies that might be perceived as only 'for the boys'. In Japan and Scandanavia, house-builders have introduced virtual reality for marketing purposes by using it as one medium among many.

Revealing hidden structure

Representations are used in problem solving to reform the problem domain and reveal the hidden structure of the

2.4 Sekisui House customer care centre, Japan

In Japan, the house-builder Sekisui House uses computer graphics to discuss the proposed design of new housing with customers. The technology is not used in isolation, but is part of a process in which many forms of representations are utilized to build up an understanding of housing design.

On arrival at the customer care centre of the Japanese house-builder Sekisui House, customers are taken to an orientation zone in which there are displays about the history of housing and the different kinds of housing available in different parts of the world. As customers pass through this and the other zones, which look at housing quality and environment, structure, lifetime home, storage, equipment, space, kitchen and interior and exterior co-ordination, they slowly build up a detailed picture of their new house. This may take place over more than one visit. There are many different types of representations, interactive displays and models that allow the house to be explored at different levels. For example, in the storage zone, it is possible to move full-scale storage units up and down the wall to determine the most suitable height. In the kitchen zone it is possible to move units around in a 1:200 scale model. As the customer makes design choices these are entered into the CAD package.

In the final zone of the customer care centre, virtual reality is used to present the whole house to the customer. Thus, virtual reality is used as part of a larger narrative about the customization of the design. It is also used at a stage where its use builds on previous discussion between the customer and the sales representative.

data. In this chapter we have seen that virtual reality is not the same as reality. However, it can be useful as a representation, for representing 3D spatial data and temporal data, despite (or because of) its differences.

The effective use of virtual reality is task-dependent and contingent upon the relation between the virtual model, its environment and users and the reality it describes. Learning and experience are important factors. It is clear that expert users of virtual reality are more sophisticated than novices. Models created can be tailored and different viewing perspectives can be used to aid novices gain spatial knowledge through virtual reality and understand the landmarks, routes and overall plan of an environment. The interface to the VR model can be used to improve performance, but the type of aids and tools that will be helpful to the user will be different for different tasks. Application is something that will always need to be borne in mind as architects and engineers attempt to use virtual environments.

As a spatial and temporal medium, virtual reality is one of a range of representations that can be used for problem solving. Though often advocated as a substitute to physical modelling, here it is argued that the most successful implementations of virtual reality are those where it is used alongside representations in other media. In this chapter it has been shown that virtual reality promises to be a powerful medium for the representation of the built environment and the exploration of potential changes to it. In the next chapter we will consider some practical uses of virtual reality by professionals within the project team.

3 Building prototypes

Virtual reality is being used in the design and construction of large complex buildings such as airports, hospitals, research laboratories and shopping malls. It is also being used on infrastructure projects, including road and railway networks. In this chapter we explore the business drivers for these uses. We look at how virtual reality enables professionals to prototype the product and 'informate' the process.

The production of large complex products, such as modern buildings and infrastructure, is markedly different from the mass production of consumer products (Hobday, 1996; Gann, 2000). Production is project-based and innovation often occurs at the boundaries between different traditional roles (von Hippel, 1988; Hobday, 1996). Virtual reality is beginning to be used in many sectors that produce large complex product systems. It is being used to co-ordinate the work of different professionals involved in projects and to improve their understanding and use of underlying engineering data. Users in the construction sector are inspired by advanced applications in oil and gas, aerospace and manufacturing.

Within the construction sector, virtual reality has often been described as most useful for the architect, allowing them to walk the client around a new building before it is built. However, it is consultant engineers, construction contractors, property owners and facilities managers that are the lead users. Virtual reality is being used within the professional project team and supply chain to visualize and manage increasingly complex engineering and design data. The lead users do not see virtual reality as a subject of interest in itself. Instead they are concerned with reducing risk, increasing technological innovation and improving business processes. Visualization is seen as a means rather than an end. One manager said 'It's not about a pretty picture'.

The major business drivers for the use of virtual reality identified by the professionals interviewed are:

- *simulating dynamic operation* – for example, to improve product quality and safety of operation;
- *co-ordinating detail design* – for example, to reduce the cost of errors and redesign work; and
- *scheduling construction* – for example, to reduce lead times, incompatibilities on site and waste.

By enabling professionals to visualize available engineering and design data, virtual reality can be used to prototype designs and consider different alternatives. One aim is to find faults earlier, when they are less expensive to correct, and another is to explore completely new solutions. At the later design and construction stages the use of virtual reality is seen as a way of reducing redesign work and construction delays. It facilitates concurrent engineering processes and is being used to increase the quality of the end product.

Professionals in the project team and supply chain are looking to use information technologies not only to automate existing processes, but also to 'informate' (Zuboff, 1988) these processes, making them visible and understandable to everyone within the organization or project team. Virtual reality can be seen as augmenting and extending the potential of CAD packages. Virtual reality models act as boundary objects (Star, 1989) around which different professionals can discuss design issues.

We will look at some examples of leading industrial applications of virtual reality for each of these purposes, and the lessons that have been learnt by the construction professionals involved.

Simulating dynamic operation

A major business driver for the use of virtual reality is the simulation of different operational conditions. Using virtual reality, the dynamic processes that buildings and infrastructure support can be explored at the design stage.

Experimenting and testing can be seen as a process of scanning or searching through alternative options. Thomke (1998b) argues that computer modelling can reduce the time and cost required for experimentation, and can significantly improve the learning derived by increasing the number of experimental iterations. This is particularly true in the design of complex products, where alternative testing strategies are time-consuming and costly. By exploring the operation of design alternatives, the quality of the final product can be improved.

Virtual reality uses technologies that were developed by the aerospace industry for flight simulation. Aeroplanes are complex products, and flight simulators allow their operation to be simulated and tested by pilots. Data gathered is used at the design stage, as well as to train pilots to fly different types of planes. Simulating and testing the dynamic operation of buildings promises similar benefits.

Many activities that must be considered in the design of the built environment are already being simulated. For example, the project management and engineering company Bechtel is testing the environmental performance of airports, using virtual reality for sound simulation. Work with a VR model was conducted for Atlanta airport, which was having a commuter airport runway added to it. The team involved in this project met in the nearby Sheraton hotel so they included this in the model as well to facilitate their understanding. The model was used to visualize noise contours, and was generated from engineering data. It took two weeks to create the model.

3.1
Bechtel model of Atlanta airport showing noise contours

3.2
Bechtel model of Atlanta airport showing noise contours

Virtual reality is also being used to simulate and test other conditions that involve the dynamic operation of buildings and infrastructures. These conditions include the movement of people, the movement of transportation and the operation logistics of supermarkets and factories.

People movement

Engineers in the engineering consultancy Mott MacDonald had an early interest in the use of virtual reality to understand people flow. They experimented with off-the-shelf commercial VR programmes and, having found that these did not do what they wanted, they decided to develop their own software. They have developed a PC-based program to analyse fire egress and people flow called STEPS – Simulation of Transient Evacuation and Pedestrian movements. Mott MacDonald has been able to market and sell its specialist expertise in crowd simulation (Plate 7).

In models visualized within the STEPS software, individual persons have pre-set characteristics. Crowds interact according to the characteristics of the individuals of which they are composed. The software can be used to analyse people flow through office blocks, sports stadia, shopping malls and underground stations – any areas where there is a need to ensure uncomplicated transitions in normal operation and rapid evacuation in the event of an emergency. Simulation and optimization of people flow can be used to provide a more agreeable environment and a more effective fire safety design in large and busy places. A key lesson is that use of 3D provides an effective tool for visualizing and exploring the process. An engineer involved in people flow pointed out that it was much easier to track and understand an individual person's movement in 3D than to follow a dot, representing that person in 2D.

Vehicle movement

Virtual reality has been used to show vehicle movement in the design of both road and rail infrastructure. Early work was conducted for the UK Highway Agency in 1997 and focused on road alignments. In the USA, MultiGen-Paradigm developed a visual model of a 2.5-mile section of Los Angeles 710 Freeway in 2000 to enable planners in the California Department of Transportation to understand the impact of development and retrofit work before construction began. The visualization contains existing conditions for the freeway segment, as well as the proposed beautification and retrofit elements. It allowed the planners to get immediate feedback of appearance, scale and compatibility of the proposed freeway improvements (Plate 8).

Interest in the accurate simulation of vehicle movement has grown with increasing concerns about safety. In the UK, there has been increased spending on the rail infrastructure after some well-known incidents such as the Ladbroke Grove rail crash in 1999. At the time of this crash, the consultant engineering company WS Atkins had several years of experience in applying virtual reality to highway projects. Using virtual reality to assess whether train drivers could see signals seemed an obvious application and within a week of the crash WS Atkins Rail was considering this technology. Collaboration between the road and rail branches of the company brought about VRail as a 'proof of concept' tool.

VRail is a virtual reality tool that gives an accurate simulation of the driver's view. Most tools that simulate the movement of trains assume that the driver's eye follows the centreline of the track, with a certain sideways offset from it. However, this is not strictly accurate. Railway locomotives and coaches are mounted on two 'trolleys' called bogies. The driver's seat is part of the vehicle and is located in front of the leading bogie. It therefore swings further out on bends; the precise distance is determined by the geometry of the track and the rail vehicle. Whereas most systems interpolate the position along the alignment, VRail calculates it accurately, taking into account the track details and the type of vehicle.

3.1 Proof House Junction, UK

WS Atkins VR Rail tool was used by Atkins Rail to check signal visibility at Proof House Junction near Birmingham, UK. As a part of the West Coast Main Line modernization, this junction was the subject of a major remodelling in late summer 2000, with the aim of reducing journey times by eliminating some conflicting routes. Atkins Rail was working on the project together with a construction company, Carillion, and the railway track management company, Railtrack. Using design and survey data, a working VR model of the junction was created within six weeks and has been further enhanced several times since.

On a desktop computer the model runs trains smoothly at true speed, producing upwards of 15 frames per second. Route selection and signal displays are set through a virtual signal box, and signals and points can be changed at the click of a mouse. The segments of track are intelligent objects that can be traversed in either direction and they carry design speed data. The status bar shows a continuous readout of distance along the route, speed, next signal name and the time in seconds until it is reached. The trains can be stopped or reversed, or their speed can be scaled down to give precise timings.

Overhead Line Electrification equipment presents one of the main sources of obstruction to signal visibility. For realism the model includes many gantries, hanger assemblies, catenary cables and contact cables. The view can be zoomed in or out, so that it is easy to say whether a signal is visible or not despite the screen resolution, which is still far inferior to the human eye. If a signal is obscured, say by a gantry leg, the system allows the operator to drag it vertically and sideways, and reports the new offsets to the status bar. The visibility can be rechecked in seconds.

3.3
View along the track, in WS Atkins' model of Proof House Junction near Birmingham, UK in Vail

3.4
View from the train driver's seat, in the WS Atkins' model of Proof House Junction near Birmingham, UK in VRail

Operational logistics
Use of virtual reality is being explored for both supermarket and factory layout and logistics. Software developers are developing solutions that cater specifically for these applications.

The UK supermarket chain Sainsbury's was one of the first retailers to look at virtual reality. In the space planning software developed for them in 1993 Sainsbury's representatives were free to explore store internals, and could pick up 3D objects by inserting their virtual hand into any of the products on display and depressing a button on their hand controller. Many major retailers are now exploring virtual reality.

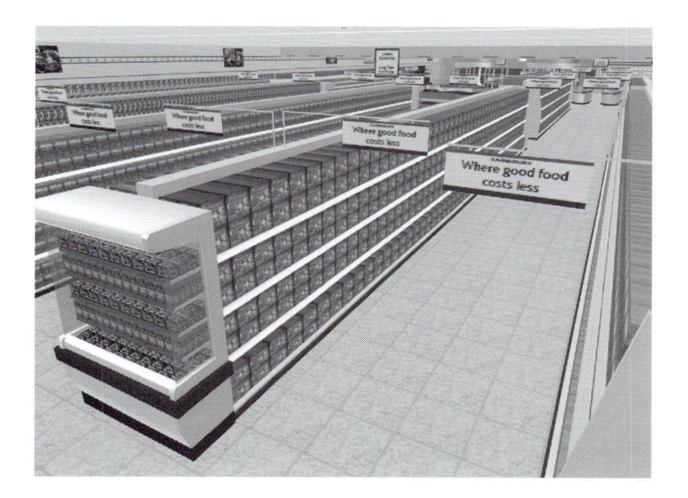

3.5
An early space planning tool, developed by Virtual Presence and used for retail by Sainsbury's supermarkets

Large automated warehouses have audit trails, so that every time an action is taken the worker swipes a bar code so that every activity is recorded. In many factories, these records are archived but the archives are not used. One VR supplier explained:

> Those archives get dumped on zip disks and put in the drawers because they just have no way to understand them. Printing them out doesn't do any good because you need to look at trends, what's happening, you need to be able to look at the big picture.

This supplier demonstrated how the data could be visualized using virtual reality, by taking one such archive and generating bar charts on the warehouse floor to represent the level of activity.

This allows the management at large automated factories to visualize their archive data and analyse how the shop floor is being used over a period of days and weeks, enabling them to identify bottlenecks and blockages in the production line. The supplier put it 'At my desk I can see if Larry is taking a smoke break'. The manager is able to access and visualize information about the process. The supplier said:

> ... and you don't have to call your IT department, say, 'Hey, I want to query this database', and wait 24 hours for the results. We can visualize them right here.

These examples suggest that virtual reality can be used not only to automate existing processes, but also to informate these processes.

Co-ordinating detail design

Co-ordinating the design of engineering systems at the detail design stage is one of the major business drivers for the use of virtual reality. The cost of making design changes increases dramatically once a project is under construction. By using virtual reality to check for design errors and incompatibilities before this stage, the amount of time, materials and money wasted on site can be reduced, lowering the overall design and construction costs. Virtual reality can also be used to improve the robustness and safety of the overall design, reducing the risk of design faults and hence the risk of litigation due to operational failures.

Construction professionals are learning from users in other sectors. For example, Jaguar Racing, a company that designs and produces Formula One racing cars, uses a 3D model to optimize design time and to identify design conflicts early in the process (Nevey, 2001). These types of benefits are being sought in the construction sector through the increased use of an interest in object-oriented techniques, product modelling. The aerospace industry was also an early investor in virtual reality, and the manufacturing company Rolls-Royce has used it in the development of its Trent 800 aero engine.

3.6
Rolls Royce Trent 800 Engine

Virtual reality was used in a petrochemical plant design for ICI and Fluor Daniel. In 1993 both companies were interested in virtual reality, not only as a complementary technology to CAD and as a means of replacing costly scale plant models, but also as a mechanism for improving working practices and reducing plant design and total life cycle costs.

3.7
ICI/Fluor Daniel Petrochemical plant project

Whilst many of the construction professionals working on fixed price projects are using virtual reality after bidding, owners may also see the provision of 3D information as a way of reducing their costs. Three-dimensional laser scanning was used to obtain accurate information about the Forcados Crude Loading Platform before a major upgrade in 1999, as there was a lack of detailed and accurate as-built drawings of the platform. As the owner of the facility, Shell Petroleum Development Company of Nigeria (SPDC) felt that sufficient communication of the work scope to the contractor teams would reduce the risk element during bidding and result in commercially attractive bids.

3.8
Two views of the Forcados Crude Loading Platform. This model has been input into virtual reality using 3D laser scanning techniques

Within construction, there has traditionally been insufficient flow of information between members of the project team, suppliers and manufacturers. Yet shared understanding of design is important to ensure construction projects fit together. A lack of co-ordination leads to constructability problems, delays on the construction site, waste and a lack of safety. There is a need for more information to be accessible to all professionals involved in design.

Though most of the professionals involved now use computers, different professionals favour different CAD packages to support their specialized tasks. The management and co-ordination of their activities, within the individual organization and in the project teams that span the boundaries of organizations, and the exchange of data between them is not an easy task. Using virtual reality to identify errors and clashes may improve constructability and reduce costly redesign work and waste. Using virtual reality to check the location of equipment and key safety controls, as elements within a number of design subsystems, may improve safety. We will look in detail at these motivations for using virtual reality.

Identifying errors and clashes

Identifying errors and clashes early in the process is important to construction contractors, as they are being made increasingly responsible for spatial co-ordination of detailed design. On many fixed price projects they are responsible for any rework that is necessary on site because of incompatibility problems. This can be an expensive process, as there are often significant costs associated with the resultant redesign and delays. As contractors often operate within 1 per cent profit margins, the management of the associated risks is a major business concern for them. Accurate construction information is vital for co-ordination of spatial layout and a manager from Laing Construction argues that 'When we accept bad quality information we accept risk'.

Slight differences between the design drawings of different professionals may lead to buildability problems when standard drawing procedures are followed. Plates 9 and 10 show such errors. In Plate 9 the two superimposed drawings show walls in different places, and in Plate 10 the door swing is different. The contractor does not know what to build when faced with these types of conflicting design details. In the past contractors employed staff to manually check through paper drawings; however, as CAD can be infinitely precise, digital drawings often go unchecked as everyone assumes that the information is right. This type of error is often introduced into the process as different professionals redraw, rather than reuse, CAD data. For the contractor, this can make the process of co-ordination worse than when it was done on paper. In the example given, the problem of co-ordination was compounded as there was poor communication and the process was not logged. Two different design professionals were alerted to the design conflict and both professionals made changes to their drawings, leading to a design that was incompatible in a different way.

For Laing Construction, the lesson to be learnt from such experience is that there needs to be a single project model to co-ordinate all data relating to a project. The company has collaborated with Salford University in work on integrated project databases to manage all project data (Aouad *et al.*, 1997). The data then needs to be visualized and an interactive 3D design review tool is used as the interface to a single project model. For Laing, visualization in an interactive, spatial, real-time medium is central to the process of ensuring accurate construction information. The NavisWorks software has been used on a number of projects including Basingstoke Festival Place, UK.

3.2 Basingstoke Festival Place, UK

Laing Construction used more than 100 CAD models for the detail design and construction of the shopping mall, Basingstoke Festival Place. Once imported into the i3D review software, NavisWorks, the in-built compression reduced the file size, allowing the entire project to be navigated in real-time (Plates 11 and 12).

Sectioning, annotation and clash detection functions in the review tool were used to highlight conflicts, including insufficient clearances for building use, maintenance or construction. For example, in one of the brick stair towers in the centre of the scheme it was noticed that there would have been no room to put up scaffolding and the design was changed. A manager said:

> *If we had not been able to identify this error prior to the building phase the project would have been subject to a significant delay resulting in huge costs in terms of time and money.*

One contractor argued that virtual reality was so valuable for them, in terms of enabling greater co-ordination of detailed design (and hence reducing the number of change-orders) that it was worth them building this 3D model for design review at the detailed design stage. Using it they could simultaneously view the CAD models from different consultant designers and fabricators and rapidly identify any problems with the constructability of the design.

Laing Construction is not alone in using virtual reality for this type of application. Another lead user is the project management and engineering group Bechtel. This company has a track record of using visualization techniques within the oil and gas sector. Though the particular software packages used are different in the different sectors, in-house skills and a reputation for using these techniques have helped the company become a lead user of virtual reality within the construction sector. Bechtel London Visual Technology Group used virtual reality to help engineers co-ordinate design and construction on the Luton Airport project in the UK. On this project there were about 10 000 construction drawings, so the potential for errors was high, and there was a need for details to be rigorously checked.

Problems of constructability can occur on site as well as in the design office. A VR model of Luton Airport showing the different engineering subsystems, such as the heating, ventilation and air conditioning (HVAC) subsystem and the steel subsystem, was put onto CD-ROM and sent out to the engineers on site. This allowed site engineers, as well as design engineers, to use the model for visual clash detection.

Using virtual reality for clash detection is seen as a money saver for companies such as Laing Construction and Bechtel. These companies may be liable for the cost of errors on site, and can reduce the risk of errors through the use of virtual reality.

Checking the location of safety-critical parts

On some projects virtual reality is being used to ensure that safety-critical parts of the design will work within the context of the total design. This is a motivation for the use of virtual reality in the design review of major rail initiatives in the UK. Like WS Atkins, the company Bechtel has become interested in using virtual reality in the rail sector. Bechtel Visual Technology Group is using a tool developed in conjunction with the engineering software specialist Infrasoft. The use of virtual reality allows engineers to model the position of new signalling equipment with a driver's eye view to help avoid red lights being missed. Rail projects on which it is being used include Thameslink 2000, the improved train line through London from Brighton to Bedford. There is a significant investment in this software, with one to two people working on the model over the lifetime of the project (Plates 13 and 14).

The VR model is being used in project meetings to allow professionals from the client Railtrack and from suppliers, regulators and consultants to review the design and ensure that safety-critical aspects, such as the signalling, are co-ordinated with the rest of the subsystems. As the data is visualized and used by all companies that are within the consortia working on the project, issues of intellectual property rights may be expected to arise, but the shared visualization of data benefits all of the companies involved.

Signals can first be appraised using the VR software. Previously, a Signal Siting Committee, comprising as many as twelve people from rail companies and safety inspectors, would make four or five visits to track for every signal layout change (Glick, 2001). The new software can reduce this to a single visit per signal, according to the Thameslink 2000 modelling manager for the engineering group Bechtel (Plates 15 and 16).

For all the companies involved in railway design, safety is a critical issue. Companies need to reduce the risk of design errors that may lead to accidents and charges of criminal negligence. Railtrack claims that their £150 000 (~US$220 000) investment in VR software is increasing safety, speeding up track improvements and saving millions of pounds (Glick, 2001). Once created, the model continues to yield value as it is reused by the signal supplier to test reflectivity of new designs and to improve signal design, and by train operating companies to train the drivers and familiarize them with the route. Checking the location of safety-critical parts reduces the risk of design errors and brings major benefits to the companies involved.

Scheduling construction

Construction scheduling has been identified as a major business driver for the use of virtual reality. There is increasing interest in its application for scheduling, which is often referred to as 4D-CAD. The use of virtual reality can reduce delays on site by ensuring that two trades are not using the same part of the site at the same time, thus reducing the length of the construction process. A US-based supplier describes how 4D-CAD can have real economic benefits on fixed price projects by improving the process. This supplier argues that on a major highway design, it was possible to reduce the construction process by two weeks in a 100-week schedule. This represents a major saving as the contracting company had 50 subcontractors working on the project, each with a daily 'burn' rate of US$50–1000 per day. However, the construction of 4D-CAD models is labour intensive, and the use of virtual reality requires high skills and high investment. It is the large companies that can afford to spend capital to recoup money on reduced operational overheads that stand to gain from the use of 4D-CAD.

Disney Imagineering Research and Development used a 4D-CAD model on the construction of the Paradise Pier

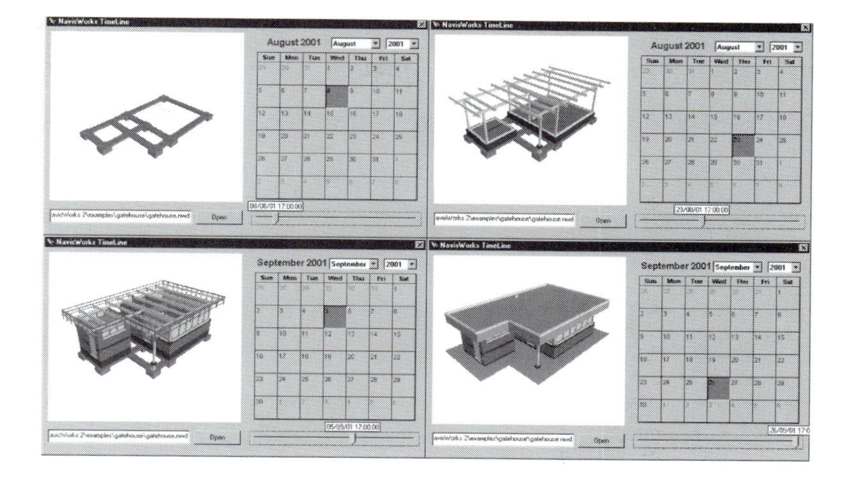

3.9
Packages such as NavisWorks can be used to dynamically assess construction sequences. This image is of a gatehouse project by the engineering consultancy Taylor Woodrow

project. Disney Imagineering worked with Stanford University on this project (Bonsang and Fischer, 2000) and felt that Stanford modelling expertise was a good fit with their expertise in high-end visualization solutions and financial rigour. Paradise Pier covers one-third of the total Disneyland Park at Anaheim, California, and was opened in January 2001. The 4D-CAD package enabled CAD data to be linked with scheduling information and viewed in a real-time environment.

Virtual Reality Modelling Language was used, and the model was created by first importing geometric data from the CAD package. This geometric data was then linked with process data for each activity, which was obtained from scheduling packages such as Primavera. These links were made manually, as the process involved making complex decisions that could not be automated. Activity types need to be created and planners and architects needed to agree on a terminology for the different activities. The 4D-CAD model used in the construction of Paradise Pier consists of about 500 000 polygons, 380 CAD shapes and 1000 links to the schedule.

The model allowed Disney Imagineering Research and Development to save money and orchestrate manpower as it enables the general contractor to say things such as 'I can't put this in here as that is in the way'. As well as simple clash detection the model allowed visual analysis of the suitability of lay down areas, which are temporary areas used to prepare the next job and can cause conflicts. Disney Imagineering used a large screen to display the models. They felt that a key lesson was that presence within a virtual environment was useful for creating connections between people, rather than just enhancing understanding of the data. One engineer summed this up as 'Problems found together are solved together'.

Disney Imagineering Research and Development looked at 4D-CAD as a tool to improve engineering sensibility and construction management. The Walt Disney Company designs, builds, owns and operates a vast amount of real estate across the globe, in the form of various Disneyland theme parks and other retail outlets. Disney Imagineering Research and Development was interested in what made most sense for Disney companies, and they found 4D-CAD a good match with their needs. The tool provides an interactive visualization in which spatial and temporal data regarding the construction process are linked. They felt that by investing capital in this they could save on construction costs.

Disney Imagineering Research and Development is interested in using a simulation-based approach to design. Eventually they would like tools that have embedded within them an appreciation of the physical engineering process, and that look at life cycle costing and building performance issues. In this regard they find 4D-CAD a good place to tackle quantitative analytical approach to construction. As they own and operate the buildings they can choose to spend more in capital at the design and build stages rather than spending on running costs in operation. They want to connect data in a meaningful way and link this back to design choices. They would like to measure and monitor everything, achieving a demand-driven approach to maintenance, by knowing details such as how many people have walked on this carpet. A longer-term goal would be to use virtual reality to visualize this type of life cycle information.

The 4D-CAD tool that has been developed in conjunction with Stanford University is being spun out of Disney Imagineering Research and Development into a separate company that will license or sell the software to other construction companies.

Drivers, barriers and issues

In this chapter we have seen that virtual reality can be used as an interface to data. Again and again the professionals interviewed stressed that virtual reality is not of interest to them in terms of aesthetics, but in terms of the access it gives them to the engineering data. It is the data shown in virtual reality that is useful to professionals that work on major projects such as airports, retail spaces, automated factories and transportation infrastructures. Virtual reality allows the spatial and temporal complexity of this data to be visualized and understood in an intuitive manner.

Virtual reality is being used alongside a range of other digital technologies and the data that is visualized comes from construction scheduling, CAD and engineering simulation tools. Some of the professionals interviewed argue that the confluence of these digital technologies may change the way that the construction sector is set up. The drivers for virtual reality are concerned with prototyping the product – testing ideas, verifying attributes and appraising options. They are also concerned with simulating the processes of its operation and construction. These drivers raise the question of the extent to which simulation of the product and process can be integrated. At present, many processes within the construction sector are based on the premise that there is a lack of information in construction as every building is a one-off and there are no prototypes. Digital models change things by providing professionals with prototypes of buildings and infrastructure before they are built.

Unique characteristics of virtual reality may enhance the potential of the company by increasing its capacity to experiment, involving more people in the innovation process and capturing ideas generated in that process (Watts *et al.*, 1998). However, there are issues and barriers to the introduction and use of the medium. These include the time taken to create models and the danger of virtual reality being viewed as just an image, rather than a prototype that can be interacted with and changed. In this section we will summarize the business drivers, and explore the barriers and issues regarding the use of virtual reality by professionals in the construction sector.

Business drivers for professional use

Major business drivers for the use of virtual reality by professionals include simulating dynamic operation, co-ordinating detail design and scheduling construction. The use of virtual reality allows risk and cost to be reduced because of faster design implementation. It also enables better utilization of large buildings and infrastructures.

Interactive, spatial, real-time tools can be seen as 'informating' processes. The companies that are lead users are typically those with the most risk that can be mitigated, or with the most to gain from improved processes. Consultant engineers are using virtual reality to improve their reputation for technical expertise. Construction contractors and project managers are using it to reduce the risk of co-ordinating spatial layout and hence increase their profit margins. Real-estate owners are using it as they can afford to spend in capital in order to save on construction costs. Facilities managers are using it to optimize the use of their facilities by understanding the operational logistics.

Issues

The data used in virtual reality may be off-line data, as data exchange between other engineering applications and VR applications remains problematic. For many construction companies, models need to be put into virtual reality and viewed quickly for the medium to be useful. This is possible; however, at present, many models take a long time to build and optimize even for small projects. As CAD models do not contain all the required temporal data and scheduling packages do not contain all the required spatial data, there may be a need to build links between the packages and this can be a time-consuming process. Issues raised include the management of the use of virtual reality within the organization and the use of virtual reality as a prototype rather than an image.

When new technologies are introduced management can be left out of the loop. The technical director of Jaguar Racing pointed out that when the company, which designs cars for Grand Prix racing, used to work on drawing boards he would walk down through the design office at night and know what was going on. If he had any concerns with the design that had been done during the day he could leave notes on desks. When the company first moved to using CAD/CAM for 100 per cent of its design work, he walked through the office and all he saw were blank screens (Nevey, 2001).

If VR models are made available to everyone within the organizations involved in a project, then they stand a better chance of being used across the different functions and processes within these organizations. One supplier argued that:

> The applications that we use are not for the back-room engineers. What I am going to show you is more for the corner office, for the management team.

Virtual reality can be used to enable discussion of design between disciplines and between engineers and management, within multi-disciplinary organizations such as large consultant engineers and commercial developers. Yet models are often not freely available and interaction with them is limited, as sharing data in an unmanaged way may lead to problems of version control. Virtual reality models are often kept and maintained by particular technical individuals within the organization. When this happens, the use of virtual reality may become separated from decision-making on projects. There is a danger that the prototyping facilities within an organization can become an innovation ghetto (Schrage, 2000).

If virtual reality is used in a limited way its use may merely automate existing processes. Members of the organization may see no benefits to the use of the technology and become concerned that it may be used to undermine their position within the organization. To gain additional value from virtual reality it must be used to informate processes and add information to tasks, enabling professionals to do them better.

For virtual reality to allow insights into the real designs behind the representations, design data must be accessible through the visualization. The more polished a model is, the more likely it is to be used simply as a vehicle for persuasion and public relations, with significant energy diverted from design to polishing and production (Erickson, 1995). There is a danger that successful prototypes become enshrined within an organization and get enhanced to a point at which they are no longer useful and just form a drain on resources. A manager is reported to have explained this by saying:

> The moment you successfully demonstrate a model can work you have people trying to add features to make the prototype better. It becomes a Christmas tree. (Schrage, 2000: 144).

Professionals will not use these models in ways that may undermine their position within the project team, even if they can see that it would be beneficial to the outcomes of the project to do so. Schrage asks the question:

> 'Would a design team be rewarded in the real world for figuring out how to creatively eliminate their sub-assembly from the prototype? (2000: 158).

The issue of whether virtual reality can be used to capture and explore issues of 'delight' within the real world is still open to question. In this chapter we have seen that virtual reality has the potential to be used to mask out irrelevant features or 'white noise', allowing the user to reveal and examine underlying structures. Yet understanding what data are shown in virtual reality requires the ability to interpret the data, understanding the limitations of the medium and the boundary assumptions. With less critical audiences, there is the potential for the deliberate and accidental communication of misinformation as well as information. Issues raised by this will be explored further in the next chapter, which looks at the use of virtual reality for generic design activities and wider communications outside the project team.

4 Design and wider involvement

The use of virtual reality changes the way we learn about space and the way we communicate our insights to others. It promises to facilitate wider understanding of the built environment and to enable clients, managers and end-users to contribute their experience. In this chapter we will look at virtual reality for design and for wider involvement in design.

It is the design of virtual spaces rather than physical buildings that is most exciting for many professionals. Using virtual reality gives designers access to new markets for their spatial expertise, allowing them to explore spatial concepts and evolve a new understanding of space in the virtual realm. Important work is being done in this area. However, the built environment is an aspect of social life (Hillier, 1996) as it is shaped by our social interactions and it, in turn, shapes those interactions. Design of abstract space, without concern for physical site or inhabitation constraints, may shift attention away from design skills and tools that are relevant to the built environment.

To be actively involved in the design process, clients, managers and end-users need to be able to understand the possibilities. Virtual reality is one medium that can be used to include them, by showing options and allowing dynamic changes to be made to design proposals. Clients are enthusiastic about its potential. An architectural journalist argues that:

> Walking through a virtual building or zooming into any nook and cranny is a lot more useful than taking one of those roller-coaster 'fly-throughs' that make you feel sea sick as you watch them inside some developer's executive suite (Glancey, 2001).

Virtual reality is beginning to be used for communications outside the project team, for example at the customer interface. Leading industrial users include the designers of large and complex buildings and infrastructures discussed in the last chapter and designers that reuse design elements on many smaller projects. Organizations working on both large and small projects are interested in improving participation in design, raising the profile of their organizations and increasing their profit margins. The major business drivers for virtual reality identified by the professionals interviewed are:

- *demonstrating technical competence* – to market the skills of the organization;
- *design review* – to improve the quality of the product and reduce risk; and
- *marketing* – to sell products and services.

The use of virtual reality for design and wider involvement is not widely diffused through the industry. Professionals in some organizations do not feel that they can obtain sufficient business benefit to use virtual reality in the design of the built environment. Three-dimensional CAD packages, rendered still images and animations are in more established use in practice.

This chapter looks at the new market opportunities opened up by virtual reality and at the business drivers that are motivating lead users – demonstrating technical competence, design review and marketing. It also considers whether virtual reality can be used for individual creative expression in the design of the built environment or whether it is 'just' useful for presentation of design. In conclusion, the drivers, barriers and related issues are summarized.

New markets

Some designers see virtual reality opening up new markets for their architectural design skills, as dynamic and spatial media are incorporated into the built environment, and as spaces are designed and represented virtually in interactive, spatial, real-time media. These designers are not using virtual reality in the traditional architectural design process and they are sometimes not designing physical buildings at all. Instead they are becoming interested in media-rich environments and virtual representations of space.

Media-rich environments

The incorporation of dynamic and spatial media into the built environment is becoming pervasive. For example, in Times Square, New York, building exteriors have been turned into enormous television screens and news-feeds deliver real-time news bulletins throughout the day (Plate 17). The convergence and overlap of digital media and urban spaces is a new challenge to designers.

Representations have been used to augment places and to create the illusion of greater space throughout history. Examples include the frescos used in ancient Rome and the *trompe l'œil* of Baroque architecture. Media-rich environments such as Times Square are a part of this tradition. Such environments give the illusion of being a kind of hybrid space between the real and the virtual.

Media-rich environments are extremely exciting to many architects. Rashid of Asymptote Architecture explains how, at the time of the Los Angeles Gateway project, his interest was in imaging, using virtual reality to represent and communicate design ideas. That interest was lost pretty quickly and he became interested in:

> Exploring virtual worlds, different interfaces, different ways that architects can explore spatiality ... my take on it was to totally radicalize the territory in which we study spatial problems.

Architects and installation artists are using media-rich environments to experiment with spatial concepts. The architectural company Oosterhuis used immersive virtual reality at the Biennale 2000 exhibition to evoke the feeling of being inside an active architectural structure, which was called trans-ports. A total of 128 sensors were built into the floor and were triggered by the public, changing the shape and content of the structure. People could also access the structure through the Internet (Plates 18 and 19).

Designers have been described as working across hybrid spaces or 'coterminous territories of the real and virtual' (Zellner, 1999). Rashid of Asymptote argues that the interest in building is the same '... whether it is in pixels and wire frames, or concrete and steel – one employs the same kind of disciplines, rationales and procedures'.

The New York Stock Exchange heard about research that was being conducted on virtual environments at Asymptote.

They were interested in visualizing vast amounts of information and had previously been studying this using 2D planes or billboards in 3D space. In 1997, when they invited Asymptote to start mapping data into virtual environments, the practice used their background as architects to incorporate many of the things that are taken for granted in architectural production into the virtual environment. Following on from this project, the New York Stock Exchange asked Asymptote to design a physical environment to house the virtual environment. They were able to market their experience in the design of media-rich spaces and they designed both the representation of data in the virtual environment and the physical environment.

In media-rich environments, digital media are layered over the physical built environment. There is a relationship between virtual and physical space and a working out of design ideas in both virtual and physical media. The design of media-rich environments gives designers an opportunity to explore spatial concepts in the built environment. However, other new markets for spatial skill, such as the design of virtual space, are shifting the focus of attention away from the built environment altogether.

Virtual space

Architects are becoming interested in the use of virtual reality for the design of virtual space itself and they are exploring virtual cities and inhabited worlds. Researchers are exploring the urban geography, social functioning and architecture of avatar-based online communities (e.g., Maher *et al.*, 2000b; Schroeder *et al.*, 2001). Some designers are using their architectural skills to design virtual space in games, Websites and multi-user networked environments.

Early advocates of virtual reality envisioned an increasing need for cyberspace architects to design virtual space, arguing that:

> Schooled also along with their brethren 'real-space' architects, cyberspace architects will design electronic edifices that are fully complex, functional, unique, involving and beautiful as their physical counterparts if not more so (Benedikt, 1991: 18).

Many leading architects, designers and theoreticians take up this challenge (e.g., Frazer, 1995; Chu, 1998; Spiller, 1998). They are interested in design within a virtual or

4.1 New York Stock Exchange, USA

The virtual environment that Asymptote Architecture created for the New York Stock Exchange allowed movement in a multi-dimensional space. The designers took into consideration things such as movement, light, structure, form and time. Sketch studies were conducted in VRML and the Alias modelling environment before everything was moved to Iris Performer. The practice felt that the reason the project was successful was that, for the Stock Exchange and for them, it was seen as a new kind of built environment. It was not a kind of abstract cyberspace. Rather than being a:

> *... free floating non-horizon non-gravitation cyber-space ... it was really based on the notion of horizon and movement and camera views and movement through space and what happens to space over time...*

Thus real world constraints were used to structure the virtual representation.

Rashid related McLuhan's axiom that when moving from one medium to another, quotation of the old medium allows an entrée into the new (McLuhan, 1964). Using real world constraints overcame the problems with a previous proposal for the virtual environment, which had no sense of place that staff at the Stock Exchange could relate to. For the operations staff, the information shown in the model is 'mission-critical'. They have no interest in form or aesthetics: they really need their data. Asymptote felt that the virtual environment should mimic, but not imitate their existing environment, so that staff at the Stock Exchange would know intuitively where to go in the model. On the other hand, virtual space was felt to have a different design rationale. Pure mimicry was felt to be a danger:

> *There was a pull towards that too, which was to make it look like the Stock Exchange, to texture map the walls with marble, to make the posts look like real posts and so on.*

When Asymptote were subsequently asked to create a physical environment in the New York Stock Exchange to house this virtual environment, they started looking for real materials that had a kind of kinship or affinity with the virtual materiality. They tried to 'emulate the tectonics and the spatial flux, the mood the virtual world has'. This was done using complex shapes and bent glass to create flow and break away from the orthogonal. Some shapes, such as the surface of the counter within the space, could only be designed by using computer modelling. On the double curved glass surface that runs behind all the data screens the architects specified a glass that looked to them as though it had built-in pixels. They liked the idea that the back wall has this sense of being almost a virtual projection but is in actual fact real.

alternative reality. They see themselves exploring conceptions of space at the 'leading edge of our world view' (Novak, 1996) rather than more utilitarian issues associated with physical buildings. Freed from the constraints of a physical site and client organization, they have become interested in self-generating and data-driven designs. They are interested in 'liquid architectures' algorithmically generated designs, and what is termed 'eversion' – the turning out of virtuality so that it is not just reliant on the technologies that support its existence, but cast in the physical world (Zellner, 1999).

For those concerned with the design, production and management of the built environment, the use of virtual reality purely as an alternative reality is of limited value. Business drivers for the use of virtual reality in the architectural design process – demonstrating technical competence, design review and marketing – are of greater interest. The needs of the end-user, client and manager of a building, and the social, political and economic factors that affect the inhabitation and operation of particular built spaces cannot be ignored. The conception of digital media as alternative reality opens up new market opportunities to architectural design practices, but it shifts concerns away from the built environment.

Demonstrating technical competence

Professionals identified the demonstration of technical competence as one of the major business drivers for the use of virtual reality in the design of the built environment. Virtual reality is being used with clients before a project starts as a part of the proposal, competition entry or project bid. It is being used to show previous or proposed projects and to market the design skills of the organization. Users include house-building companies and consultant engineers.

In Japan, the customer usually owns the land on which the house is to be built. At the pre-contract stage, the customer is free to work with more than one house-building company and to compare the options and services provided by each. Customers work with house-building companies to customize the design of their proposed house using a range of standard house-types and their related options. Japanese house-building companies, such as Sekisui House and Mitsui Home, are using virtual reality to market their proposals at this stage. Mitsui Home estimates that 60 per cent of VR use takes place before the customer's decision to purchase a house from the company.

The use of virtual reality allows the house-builder and customer to agree on an image of the house and it promotes the sale. Japanese house-building companies feel that virtual reality reduces the potential for problems arising due to customers having a different understanding of the project scope.

Companies that work on larger, more complex building types also feel a need to demonstrate their technical competence to potential clients when negotiating projects. Consultant engineers and project managers trade on their reputation and need to be able to demonstrate that they have a track record of successful design projects behind them. They use virtual reality to promote bids and market their company. For example, one consultant engineering organization found it useful to send copies of a CD-ROM containing models and visualizations to the client when bidding for the construction of a new overseas facility. Indeed, one visualization specialist within the organization argued that virtual reality was more important for winning work than for the design process. Demonstration of technical competence was a major driver for this organization, though virtual reality was also being used for design review.

Design review

Professionals identified design review as a major business driver for the use of virtual reality. Large repeat clients with developed experience of commissioning buildings and infrastructure are also enthusiastic about this use. Even when there has been a good briefing process, clients' needs may change and develop as consumers discover rather than know what they need and their original priorities may not be the best ones (Kodama, 1995). Design organizations are interested in using virtual reality to improve clients' understanding of design options and also to improve their own understanding of evolving client requirements. Design review can be used to fully involve clients, managers and end-users in decisions at the design stage, through a participatory design approach, or it may be used to give more limited scope to make changes.

Users evaluate the built environment differently from designers (Zimmerman and Martin, 2001). Participatory design approaches attempt to bridge this gap in understanding between users and designers. Through participatory design, experience of the built environment in operation can be fed back into the design stage. The aim is to listen to all social groups to ensure that new buildings fit the needs of their future occupants. The need to achieve equal access for disabled people has driven some later developments in participatory design. At Strathclyde University, researchers have used an immersive VR system, with a 150 × 40° screen projection and a reactive motion chair, to simulate the movement of manual wheelchair users through the built environment. When real wheelchair users use the VR system to explore new building designs, data can be obtained about potential collision points. This data can be used at the design stage to better tailor buildings to the needs of wheelchair users (Conway, 2001). In such processes designers can learn from others' experiences and benefit by capturing their knowledge and innovations within the process.

Designers have been described as resistant to participatory design practices, following technological and architectural fashions for their own sake (Derbyshire, 2001). Design review can be seen as a programme of reality checks throughout procurement to keep designers in touch with clients' constraints and to protect end-users' interests (Leaman and Bordass, 2001).

The use of virtual reality for design review and participatory design is part of a tradition that precedes the availability of 3D computer graphics on personal computers. In the 1970s, the Laboratory of Architectural Experimentation (LEA) in Lausanne simulated built space at full scale (Lawrence, 1987). The laboratory was designed to aid creativity in architecture. Lightweight plastic building blocks and moveable platforms were used to create a model that allowed exploration and experimentation with spatial forms and dimensions.

The laboratory was the venue for design studies by architecture students and also the simulation of designs for part of a co-operative housing scheme. A design-by-simulation process was developed to enable residents to mould proposed house designs according to their requirements. The professional experience of an architect was used in this process and design-by-simulation was not seen as automating design or diminishing the role of design expertise. Simulating design at full scale enabled end-users to participate in the design process and provided designers with information about their needs. Thus, design-by-simulation adds information to, or informates, the process.

A key finding of the work at the laboratory was that the prototypes used should be simple renditions of buildings that do not inhibit the development of alternative designs. They should enable design proposals to be simulated and evaluated as simply and as quickly as possible and be designed to focus attention on the size and shape of the rooms and the interrelationships between them (Lawrence, 1987).

These lessons are relevant to us when we use virtual reality for design review and participatory design. Clients and end-users will feel more comfortable making design changes if it does not appear that decisions have already been made and designs finalized. The types of models used affect the benefit obtained in design review.

4.1
Use of the Laboratory for Architectural Experimentation (LEA) during the initial stages of the design-by-simulation process

4.2
A system of prefabricated door and window frames was also used in simulation at the Laboratory for Architectural Experimentation

In industry, virtual reality is being used most widely for design review at the later design stages. Lead users argue that even late in the design process the models created offer considerable benefits for design review and that they have considerable reuse value. Where virtual reality is used for design review, professionals can gain additional benefits from the models created by using them for marketing later in the process.

The extent to which virtual reality is useful for design review depends on project characteristics, such as the complexity of the project and the level of component reuse. There is a trade-off between benefits and the time taken to create virtual reality models. The rest of this section looks at the benefits of using virtual reality for design review in a range of project types: standard or customized housing and interiors, small unique buildings and large complex buildings.

Standard and customized housing and interiors

There is an established use of virtual reality in the design review of standard and customized housing and interiors. Here, design choices are made from a limited palette of pre-determined options and a library of optimized models can be built up in virtual reality to represent these.

Similar techniques are used in consumer goods markets, such as the furniture market, where the American furniture company Office Depot has included 3D models of all its furniture on its Website. To facilitate office fit-out these virtual models can be added to a virtual room. They allow professional and amateur office designers to design from a palette of options and visualize the resultant fit-out before making large furniture purchases. The extent of design reuse facilitates the use of virtual reality at the customer interface.

In the housing sector, the Kitchen Planning Support System (KiPS) was an early system developed by Matsushita Electric (Panasonic) and it has been used for collaboration between clients and demonstrators in Matsushita

4.3
Matsushita's networked VR-supported design tool

showrooms since October 1994. It was developed to allow the customer to design a kitchen by assembling components. This VR system has been extended to an application of a networked VR-supported kitchen design system (Fukuda *et al.*, 1997) to allow customers to design at home. More recently an Internet-based application for the interior design of the whole house has been made available, and this has been marketed to house-builders, rental housing firms, carpentry companies and furniture makers.

As mentioned in the section 'New markets' above, some of the major Japanese house-builders are using virtual reality with their clients both before and after the contract is signed.

The VR facility in Mitsui Home's Tokyo offices, Mitsui Home Image Planning Square, is of particular interest because it was used in a carefully tailored narrative about design options. Many customers are novice users of virtual reality and it is important that they do not feel frustrated or incompetent at using the medium. At Mitsui Home, the customers' experience was highly supported. The sales representative sat with customers and guided them through the presentation. The operator who moved the viewpoint in response to user commands was not in the same room. Instead, the operator sat behind a small glass window in the projection room. Thus the customer had control of the presentation but was not able to get lost or become distracted by insignificant details. Not all of this house-builder's customers chose to view models in the facility but it was a service that the company offered.

House-builders such as Mitsui Home have explored the use of virtual reality on both high-end and low-end systems. Mitsui Home Image Planning Square was a high-end system, which looked similar to a boardroom with a large screen (of approximately 1.5 m × 2.5 m) mounted on one wall. The images presented were therefore at nearly actual size. The application is an early industrial example. As hardware and software improve, this type of service is increasingly being delivered to customers via low-end systems such as personal computers.

Small unique projects

In the car industry the design costs for any particular design are spread over a production run of about one million; however, some parts of the construction sector have completely different economics of design. Buildings are designed to satisfy particular client requirements within a set of constraints and opportunities associated with location, budget, etc. The modelling and visualization of unique designs must provide commercial benefits over a single project.

The use of virtual reality for design review on one-off or unique projects is less widespread than its use on projects with component reuse. Many architects feel that they would obtain insufficient business benefits from using virtual reality for design review and few of those interviewed were using it. Some feel that it is not in their interests to use it and are concerned that virtual reality may limit the scope for them to be creative and use their architectural imagination. On small unique projects designers are not able to benefit from the economies of scope associated with reusing VR models. One architect argued that:

> If your client tells you to use it then obviously you would do, but in terms of the business benefits of choosing to use it with a client, that is 'all rubbish' – when you are with a client you are telling the client a story and the story is very carefully choreographed.

The lack of an inherent narrative structure in virtual reality was seen as an issue. One concern was that virtual reality might not provide designers and their clients with the right balance of participation and control. 'The problem with using virtual reality for client review is that you can give wrong ideas so incredibly easily', argued the IT manager from one architectural organization. In this organization panoramic views were seen as more useful, as they allowed the architect much more control to pick key locations for viewing within the digital model. The opinion that there is nothing that you can not explain with 'one glance at a decent drawing' was expressed.

Advocates argue that virtual reality gives clients, managers and end-users greater ability to explore and understand design as they can view proposed buildings from the egocentric viewing perspective, through which they normally experience the built environment. The IT manager mentioned above counters such argument by saying it is naïve to think the clients should be shown what they will experience every day. For this manager, the everyday experience represents only one aspect of a building and showing this would be like giving only one chapter of a book. Using virtual reality was compared to taking the ingredients used by a great chef rather than allowing the chef to craft and create a combination to put in front of you.

These attitudes may or may not be in part attributed to a resistance to participatory design methods and a lack of experience using virtual reality for design review. The lack of intrinsic narrative structure in the medium is an issue, but virtual reality can be used in a structured manner within a wider discussion about design. Some lead users of virtual reality have set up a series of predetermined viewpoints to help guide users through a model, familiarizing them with key aspects of the design and pointing their attention towards key decisions in the design review process.

Designers of single unique buildings have particular barriers to the use of virtual reality. If these barriers can be overcome the use of virtual reality may serve as a programme of reality checks throughout the design process, allowing clients and end-users to understand design and make decisions at a stage when the costs associated with change are not prohibitive.

Large and complex projects

Virtual reality is being more widely used for design review in the design of large and complex products. Companies working on large and complex products can get additional benefit from VR models by reusing them at different stages of the design process.

In the development of other complex product systems such as cars, virtual reality is used as a prototype and allows the development and management functions to discuss design issues. The innovative interaction tool EASY2C allows managers and engineers in the German car manufacturing company BMW to rotate a physical prototype in their hands and use this as an intuitive interface through which they can interact with the virtual data displayed on a screen.

In the construction sector, virtual reality has been used for design review with clients and end-users on a number of complex project systems, such as airports and hospitals. It is seen as particularly important for value engineering, where costs need to be reduced and decisions need to be made about obtaining value for money without compromising design quality. One visualization specialist argued that clients make different choices when they can see the

4.4
The EASY2C tool used in the German car manufacturing company BMW. It was developed by RTT in co-operation with BMW and the German Aerospace and Space Agency (DLR)

impact of their decisions. They often reject the least cost solution when they can visualize it, as they can see the quality difference between proposals.

Virtual reality is being used to tailor environments to user needs. The engineering and project management company, Bechtel, described their use of virtual reality on hospitals and cancer treatment centres as enabling them to fine-tune the environment 'so that the patient is comfortable'. They are using virtual reality to check critical design issues with clients and members of their clients' organizations on a number of large projects.

For large consultant engineering and project management companies such as Bechtel, the use of realistically rendered models with clients is supplementary to the use of VR models within the project team. For example, on the London Luton Airport (LLA) project, Bechtel was contracted to work with Berkeley Capital and LLA, building and running the airport on a 30-year running deal. From the same set

4.2 Dubai International Airport, UAE

Virtual reality was used with the client and with end-users in the design review of Dubai International Airport. The external and internal signage at the airport was checked using virtual reality. This was done to ensure that all signs could be seen and that both in the vehicular entrances to the airport and within the terminal building signs were in positions where there was enough time to respond to them. The client's security personnel walked through the model to confirm that there were no problems with the proposed design from a security point of view.

For nine months, a small group of modellers at Bechtel worked in parallel with the architects and engineers on the project. They created and updated a model that was used within the professional project team and with members of the client organization. This model could be used in a multi-user avatar-based manner, with the engineers from Dubai and the modellers from San Francisco meeting virtually within the model.

The model was used to check design and view different design options. Samples of carpets and other furnishings were given to the visualization group by the architects to allow them to build these textures into the models to show to the client. The model allowed the client to make decisions about alterations to the design before the building was constructed. A number of design changes resulted from the use of the model.

4.5
Screenshots from the VR model of Dubai International Airport, showing different design options on the interior of the building

4.6

4.7

4.6–4.8
Screenshots from the VR model of Dubai International Airport, showing different design options on the interior of the building

of CAD data, two models were created in virtual reality. One was used for co-ordination of detail design (professional use) and the other for design review (wider communication). The models were quite distinct; they were located in different folders on the computer and maintained separately. The former model, for improving co-ordination of detail design showed the heating ventilation and air conditioning (HVAC) subsystem, the steel work, and the design of floors and stairways to allow clash detection to be conducted between subsystems. The latter model was purely for architectural visualization and showed surface finishes and details.

Using virtual reality on large complex projects involves significant investment. For example, the Bechtel visualization group worked on the London Luton Airport project for over a year. The significant resource invested in model creation and maintenance was felt to bring business benefits and savings at the design stage and on site. This model created for architectural visualization was shown to LLA, Berkeley and the operators, such as easyJet and First Choice. As it was separate from the engineering model it could be optimized to show the internal layout and features. The whole model was not loaded up to show the client, as there were far too many polygons. When demonstrating the model the Bechtel manager could turn off the geometry to make it run better.

The use of virtual reality for design review was seen to provide particular benefit on large complex buildings. Though significant time and money is often spent on model creation and maintenance, project budgets are larger and the potential savings are greater. Models can be reused at different stages of a project. Models created for design review may be reused for marketing the finished project or for maintaining the facilities when they are in operation.

Marketing

Marketing was identified as a major driver for the use of virtual reality by many of the organizations interviewed. Virtual reality was being used on small speculative developments, where facilities were marketed to promote sales to potential customers; and on large projects, where the completed facilities were often marketed on behalf of the client.

Virtual reality is being used by organizations working on small projects, such as speculative housing developments. Housing developers in the UK operate speculatively and face considerable risks, as they often have no known buyer at the start of the process. Being able to sell from plan is a major advantage of virtual reality for these companies as it reduces the risk of development. Often there is no suitable example of a house-type in the near vicinity to show a prospective client. Virtual reality allows a house-builder to show their house-types to prospective clients through a computer screen at any office or show-house (Whyte, 2000).

Housing developers have used immersive virtual reality to get press coverage and to sell from plan. For example, the developers Persimmon Homes marketed their prestigious new apartment blocks in Sheffield city centre by giving VR tours in a bar near the city centre. Local press covered the event and sales were promoted, though the apartments had not been constructed.

On large projects, such as banks and airports, the client may want to use virtual reality to market the completed facility. Skyscraper Digital, the visualization division of Little & Associate Architects, has worked with two major banks in the city of Charlotte in North Carolina, USA, to develop

4.9
VR model of the Persimmon Homes development created by the VR supplier Antycip UK

models of their new downtown headquarters. Virtual reality was used by each bank to see what their new headquarters building would look like on the skyline and what it would look like from different parts of downtown, from the airport, from the highway, etc. As well as using the models for zoning board approvals, town hall meetings and to communicate the plans and get approvals, the banks have used them to raise the profile of the developments, for general marketing and for publicity. They have been able to get television coverage on news bulletins.

Airports, such as Schiphol in the Netherlands and Munich in Germany, have offered a virtual tour of their facilities in order to market them to international businesses and users and to advertise rental space to potential event and office renters. A VR model of Schiphol Airport was displayed using the CAVE in an immersive VR facility. For the companies involved, SARA, Zegelaar & Onnekes and Hans van Heeswijk, the interactive viewing in an immersive VR facility was seen to extend existing technical options, potentially leading to increased market share and profitability.

Munich Airport Centre markets its facilities using an online VR model. Visitors can choose their own avatar to enter the Centre in the networked virtual environment. Using the virtual model of the facilities to advertise rental space is highly cost effective as catalogue requests from potential

4.10
Munich Airport Centre, Germany

4.11
Munich Airport Centre, Germany

renters become unnecessary. Graphic objects can be used to visualize event stage set-ups or office furnishings in an office area. The online virtual environment can be used to plan events in the roofed 10 000 m^2 event area flexibly, quickly and cost effectively. Online models of airports may be visited remotely from any part of the world, which is of considerable advantage to airport facility providers, as potential renters may be based internationally.

For the lead users, some of the advantages of virtual reality for marketing are obtained because of its novelty value. However, virtual reality may have more long-lasting benefits for both design and presentation. Lead users are well placed to develop the expertise needed to benefit from and exploit future technological developments. Current advertising focuses attention on glossy images, but virtual reality could be used to take attention away from the architectural style and show the consumer the spatial layout of new buildings, the potential for change over time and the improved running costs of efficient construction.

These business drivers, demonstration of technical competence, design review and marketing, involve the presentation of design ideas and their discussion by clients, managers and end-users. But is design useful for design generation, or 'just' for the presentation of design? In the next section we will consider the use of virtual reality for generating design in the production of the built environment.

Generating design?

Design generation and individual creative processes were not identified as major business drivers for built environment applications of virtual reality. Yet virtual reality is having a broad influence on design thinking. Henderson points out that 'there are no one-way relationships between machines, people, mental models, representations, and constructed technology' (1999: 13).

Designers are themselves learning about architecture and building through using virtual reality.

Early advocates of VR systems believed that virtual reality would be useful for design generation, arguing that it supports spatial thinking and rapid exploration of alternatives (Furness, 1987). We have seen that virtual reality can be used in the design of media-rich environments and virtual space. Yet architects do not unreservedly welcome virtual reality and few architects were found to be using it for the conceptual design of physical places. There are many potential barriers to the widespread use of virtual reality in design generation. These include inadequate support for design within the current generation of applications and an unsophisticated and inappropriate use of VR representations for design. This section looks broadly at the potential of virtual reality in the design of the built environment. We will look at the changes brought about by digital media and the role of visualization at different stages of the process.

Digital media for design

Design of the built environment is a process that is being revolutionized through the use of digital media (Mitchell and McCullough, 1995). Alongside paper-based practices, designers are increasingly using a range of overlapping digital techniques: object-oriented design, 3D scanning, 3D printing, parametric modelling and virtual reality. Understanding of design is affected by the medium used and Henderson notes that:

> Young designers trained on graphics software are developing a new visual culture tied to computer-graphics practice, that will influence the way they see and will be different from the visual culture of the paper world' (1999: 57).

Architects who have grown up with digital media are expert users of interactive, spatial, real-time environments. Whilst early users saw CAD automating existing processes, these designers are solving problems using representations that do not emulate paper-based media.

Experimenting and prototyping in digital design media has led to the creation of innovative architectural forms. The development of curvilinear forms in the architect Gehry's buildings in the 1990s, for example, is attributable to the use of the CATIA CAD package. Large architectural practices, such as Foster and Partners, and Kohn Pedersen Fox (KPF), are using parametric modelling tools to morph and play with 3D forms that could not be easily imagined outside the computer. Computer numeric controlled (CNC) machines are used alongside advanced CAD systems to create physical models from digital data, allowing designers to constantly move between digital and physical, exploring the evolving design in more than one medium.

Designers are beginning to experiment and play with the VR medium and to use it with these other digital technologies. They are demanding more technically sophisticated design packages that incorporate both the interactive, spatial, real-time medium available in VR packages and sophisticated professional design tools such as those available in CAD packages. They are also demanding better support for material qualities so that the effects of glass and light can be described in interactive, spatial, real-time environments.

Some architects argue that plans and sections are better tools for the organization of spatial structure than real-time rendered environments. Yet virtual reality is beginning to be used in conjunction with other representations in digital and physical media. We do not have to choose either one medium or the other, but are free to explore design using different forms of representation across a range of media.

Visualization and design generation

Design can be seen as an iterative process of generating and testing design ideas. The designer has been described as '... having a conversation with the drawing' (Schön, 1983). Designers use more than one method of organizing information about their designs and may shift attention between different modes of thinking – looking alternately at features such as spaces or structures (Lawson and Roberts, 1991). At the early stages few decisions have been made and ideas are imprecise, whilst at later stages the design solution is more concrete. Throughout the process, designs have to be visualized so that they can be understood and communicated.

'Seeing is believing' is how the old saying goes. Highly realistic representations, such as those presented in radiosity rendered images, photo-realistic walkthroughs and animations, are instantly recognizable to most people and they believe what they see. Though using virtual reality in a highly realistic manner may impress clients, there are concerns that its use might make it difficult to focus attention on the relevant issues at early stages and may make designs look fixed.

Pen and paper sketches and cardboard models continue to be used in early design because they support ambiguity, imprecision, incremental formalization of ideas and the rapid exploration of alternatives (Gross and Do, 1996). Designers can discover hidden features in a representation if they do not get stuck with one single interpretation of it (Suwa *et al.*, 1999). One visualization specialist argued 'If the building design is very sketchy, if the designer just has a vague idea of something then obviously you can't go to 3D because it becomes ... real'.

One company created a VR model to give the general impression of a new building that was being designed for an industrial facility. They had since been asked for the specific colour they used on the walls of the model. This paint colour had not been carefully considered, but its precise description was sent to a paint manufacturer as a Red Green Blue (RGB) value and the colour was used in the building.

To address problems of realism, researchers are developing VR applications that use simple abstract shapes and limited palettes of colours, to support design at the early design stages. In the package Sculptor, for example, the concept of the 'space element' is introduced (Kurmann *et al.*, 1997). This element consists of no material and carves out a space when it intersects with a solid element. The use of both solids and voids results in a more intuitive approach to the use of the computer as an architectural design tool at the conceptual design phase.

Commercial use of virtual reality is mainly in the later stages of design. However, here too, some designers have experimented with the use of virtual reality in a more abstract manner. The Dutch architects Prent Landman have used virtual reality for design review in their work on the Westeinde hospital in the Hague, Holland. Rather than presenting a realistically rendered impression, the VR model presents only the key design decisions that make up the scheme design. This model, which was made for the architects by the software company Mirage 3D, is successful in drawing attention to the overall massing and layout, rather than more minor details. It was used to gain the acceptance of the local community living around the hospital (Plates 20 and 21).

Virtual reality is being more widely used for testing and communication of design solutions than for generating design and it may be this part of design for which it is most useful. A range of representations can be generated in virtual reality, using egocentric and exocentric viewing perspectives and varying degrees of abstraction and realism. The use of egocentric viewing perspectives, with a viewpoint within the model, may not be appropriate for design generation. For design tasks, particularly in the early stages, representations that allow a whole problem to be seen within a single view may be better. In the past, such external views have been more widely used for design generation. Henderson argues that:

> Although renderings in perspective played a historical role and continue to generate financial and organizational support for design and commercial promotion, these are not design functions (1999: 32).

Research is being conducted to develop more sophisticated VR tools for designers that allow exocentric views of abstract models in virtual reality to be used in design alongside other virtual and physical representations.

Drivers, barriers and issues

New market opportunities, using virtual reality and other multimedia techniques, are shifting the focus of many designers away from the built environment to the more profitable emerging markets related to the design of games, Websites and networked virtual environments. Spatial concepts and new understandings of space are fed back into the design of the built environment from these fields.

According to the companies interviewed, major business drivers for the use of virtual reality in the design of the built environment include demonstration of technical competence, design review and marketing. For these tasks virtual reality is used across a wider range of projects than those considered in the last chapter. However, there are high barriers to entry and as we will discuss further in

Chapter 6, the nature of the project affects the potential to obtain business benefits from the use of virtual reality. Lead users are benefiting from economies of scope, using virtual reality repeatedly over the life of large and complex projects such as airports, or across many small projects with design reuse, such as customized housing. One visualization specialist commented that 'the more uses you can find for a 3D model the less people object to the time and cost'.

Occupation and inhabitation of space is a continuous process (Lawson, 2001). Virtual reality is being used to feed back the knowledge that clients, managers and end-users have about inhabitation into the design stage. Researchers are also studying these patterns of spatial use. The social dynamics of inhabitation are being explored in research at Sydney University using avatar-based multi-user worlds (Maher *et al.*, 2000a). Inhabitation is also being explored using a range of abstract representations, including virtual reality, to simulate and display patterns of spatial use using agents with vision (Turner *et al.*, 2001).

The way that seeing becomes equated with believing can cause difficulties when realistic representations are used, both in design and presentation. There was concern that virtual reality might show more than had actually been designed. Though Japanese house-builders felt that showing details in virtual reality reduced the possibility of litigation, one UK housing developer had experienced legal difficulties after releasing a still computer image. The image of a future development showed a hallway filled with furniture. They had been sued when they could not deliver this as this hall was an escape route and the fire officer did not allow the furniture (Whyte, 2000). With the increasing use of digital media for both design and presentation we can expect the legal situation to change, but professionals will still need to consider the extent of detail shown in presentations.

Virtual reality can be used as part of a strategy of obtaining feedback from clients and end-users, but there is no simple technological fix to getting wider involvement in design. The effect of virtual reality is ambiguous, particularly with regard to the degree of interaction or participation allowed in the design process. Clients and end-users may be shown the final building in virtual reality in a controlled manner, with no real or perceived possibility for feedback and comment. The way that the design is modelled may affect understanding of design, by focusing attention on particular features, with more abstract models allowing us to see features at a larger scale. As we become more sophisticated users of interactive, spatial, real-time software we will learn the extent to which the designer needs the control to tell a story and focus clients' and end-users' attention on relevant design issues.

As we will see in the next chapter, issues regarding public participation in design are also being explored at the urban level.

5 Revisiting the urban map

This chapter looks at how virtual reality can be used at the urban scale. Visualizing data is becoming important as more complex 3D information is collected and used to manage urban areas and plan their future development.

Generating, processing and exchanging information is important to cities' competitiveness in the global economy (Hall, 1995). Patterns of activities within cities are reforming around the shifting networks of information flow (Castells, 1989, 1996). Rather than spatial use simply deconcentrating as a result of information and communication networks, there are parallel trends of urban concentration and deconcentration. Power and skill remain concentrated in a few major international financial and business centres, such as New York, London and Tokyo, which offer a high-density of face-to-face contact and act as nodes within global flows of capital (Sikiaridi and Vogelaar, 2000).

The transportation and communication networks of the industrial and pre-industrial city provide the context for new information networks, affecting the location and routing of the optical fibre and wireless networks that form the infrastructure of the digital economy. The information-intensive cities that form nodes have disproportionate access to global transportation and communication networks, for example, New York City has the largest Internet presence of any city in the USA, accounting for 4.2 per cent of the national total (Moss and Townsend, 1997). Cities exist in networks and such global cities may have closer connections with other major centres at the international level than with their local surroundings.

Many twentieth-century cities were planned and managed in a centralized manner, with experts developing zoning and development control practices to determine the location of different activities, such as housing and industry. Within the network society, this modernist approach is inappropriate to the development of prosperous cities and urban regions. Instead of controlling and determining the use of space, planners aim to provide infrastructures that support open and flexible activity patterns (Sikiaridi and Vogelaar, 2000).

Planners have increasingly sophisticated tools for managing network infrastructures. Maps, which hold spatial data, have been discussed in Chapter 2, where they were described as 2D representations. Yet the nature of the map is undergoing a process of change. Geographic information systems are transforming the way that information about cities and urban areas is stored, managed and accessed. Using GIS, users can query spatial data-sets and represent the results in multiple ways, using both 2D and 3D interfaces. In both the public and private sectors, virtual reality is being used with CAD and GIS. Some models are offline – developed using GIS and CAD data but essentially separate from them – whilst others are more integrated with source data. These models of cities can be built and used in-house within municipalities and government departments, but most are outsourced to companies that specialize in building and maintaining urban models.

Within the public sector, motivations for using virtual reality are related to the perceived public good, rather than to the reduction of business risks and improved profit margins. One policy maker argues that visualization tools give people a sense of phenomena in physical space but allow superficial details to be peeled away so that the complex interdependencies under the surface can be seen and understood.

Municipalities and government agencies are looking to use virtual reality for urban management and planning. Businesses are also using the urban models that exist in the private sector, for infrastructure management, planning approvals and location marketing.

As urban simulation technology attracts wide usage by city authorities, industrial organizations and citizen groups, there are calls for wider government funding of its implementation. Lead users are using virtual reality to visualize data about the urban environment so that the data can be understood and used effectively in decision-making processes. Advocates argue that it is important to show how things happen and one consultant remarked:

> This is no longer a technology that belongs solely in the lab, working on a grant to grant basis because it hasn't proved itself. It is so essential in terms of infrastructure to modern decision-making, we really have to treat the investment as infrastructure.

Urban management and use

Virtual reality is beginning to be used in the management of the urban environment. The information that exists about a metropolis is hard to comprehend in its totality. This situation is analogous to that in the financial sector, where representations are being used to organize large amounts of real-time financial market information in order to enable professionals to act faster. Good representations allow rapid understanding of the relevant features of a data-set. The use of virtual reality and information visualization in the finance, entertainment and manufacturing sectors has influenced the development of tools for planning and managing cities and urban areas.

The founder of one visualization company that specializes in urban simulation came from the manufacturing sector and was initially motivated by a need to explain the functioning of the hydraulic pumps sold by the family business. The machinery was complex and its operation was not explained well in pictures or in animations. Following a contract with the city of Vienna, the fledgling company started to collaborate with municipal authorities to similarly show the workings of the city.

We have seen in Chapter 2 that people are beginning to use virtual reality to aid their navigation within the existing built environment. The development of GIS is one of the enablers for the use of virtual reality at the urban scale, providing the data that then need to be visualized. Used in combination, virtual reality and GIS can informate urban management. First we will look at the development of data-sets and then at their visualization.

CAD and GIS data

In the 1990s, universities recognized the need for flexible intuitive urban representations and began conducting research into interactive, spatial, real-time computer applications. Some of the large-scale urban CAD models, built as collective projects by students in the schools of architecture in the 1970s and 1980s, were translated into virtual reality. Examples include the virtual cities of Glasgow and Bath (Bourdakis and Day, 1997; Ennis *et al.*, 1999). These were proof-of-concept models and demonstrated that virtual reality could be useful in the planning process, enabling the visualization of different planning solutions.

Typically these early models were geometrically complex and involved thousands of person-hours of work. They contain a high number of polygons because of the geometric detail within the source CAD models and are commonly split into different sections that are then optimized. The model of Bath, in the UK, for example, was created from CAD data, which were structured into different layers, translated into VRML and then optimized by adding different levels of detail. The model is sub-divided into 160 submodels with four levels of detail. It covers 2.5 km × 3.0 km and consists of well over three million polygons (Bourdakis and Day, 1997).

As input and data management techniques have improved, large-scale urban models have been built in increasing numbers to represent cities from around the world. Some model builders, such as the Urban Simulation Team at the University of California in Los Angeles (UCLA) have taken a different approach to model creation than the model makers of Bath and Glasgow. The UCLA models are not built exclusively from CAD data, instead the team has used primitive forms and texture mapping to build the models. This is an approach widely used in the flight- and driving-simulator communities and the resultant models are less geometrically complex allowing a wider urban area to be visualized and interacted with in real-time. The Urban Simulation Team also advocates links with GIS software and has worked with city planners to look at applications. The model will eventually cover the whole of the Los Angeles basin, which will be detailed down to 'the graffiti on the walls' (Ligget and Jepson, 1995; Jepson *et al.*, 1996) (Plate 22).

Recent developments in GIS are revolutionizing the way that spatial information is collected and stored in cities and urban areas. Major cities have complex and overlapping infrastructures for water, electricity, waste, gas, communication and transportation. Managing the interface between different service infrastructures is not a trivial task.

Information is often fragmented and partial, with utility and municipal departments maintaining 2D maps of their own systems. To plan a new subway, for example, the local authority and transportation companies have to rely on information across many maps to determine what obstacles the excavation and construction work will encounter. Even to plan new street lighting requires data across many maps. Though individual maps may be internally consistent there are often discrepancies between the information in different map-sets. Like the construction contractors discussed in Chapter 3, city departments and authorities

5.1 New York Base Map, USA

Within New York there has been an initiative to map the city accurately and establish a 2D base map that can be used by all of the municipal authority departments and utility companies. This initiative was first led by New York City's Department of Environmental Protection (DEP) water register. Rather than just mapping the water facilities they created a map of the entire street grid. This accurate street grid can be used as a standard on which each department can base its own mapping activities. Previously the city planning department had maps based on street centre-lines and property outlines that it used for zoning and planning. Other departments maintained separate maps, for example those used for tax purposes. All of these maps were inaccurate in different ways. One interviewee explained that everybody had lots of maps but this was the first map of the whole city that would be physically accurate.

Once the city had a base map, the authorities found that having access to integrated data had major benefits. For example, in the New York Police department, data used to be held locally so that the commander of each precinct saw the local precinct crime pattern. This meant that it was very difficult to measure crime patterns across precincts. By providing integrated maps the police found that they were able to achieve much faster reactions to shifting patterns of crime.

In New York, the base map is seen as a first step. The information in the map is 2D, yet the management of a city such as New York is essentially a 3D problem. Co-ordination of different infrastructure systems, both below the ground in the subway system and cabling networks, and above ground in the high-rise buildings, requires 3D data. As there is a desire to collect and use more 3D data, there is a need to develop techniques for visualizing and interacting with these large urban data-sets effectively.

are finding that spatial co-ordination is problematic in the absence of integrated information sources. Many city authorities, such as New York, look to computer-based technologies such as GIS and VR to help them integrate and visualize data-sets. Using GIS they can create different layers, adding different information about the city to an agreed base map.

Municipalities and large private companies are looking at the use of virtual reality for the visualization of these large urban data-sets, particularly in applications where rapid understanding of 3D data is important or where mistakes are costly. Virtual reality promises to allow professionals to visually organize complexes of interconnected spatial information. Two strong drivers for the use of virtual reality are emergency response and infrastructure management.

Emergency response

Emergency response is a major driver for the use of virtual reality at the urban scale. The importance of interactive, spatial, real-time visualization for emergency response was emphasized by professionals in the USA in interviews that were conducted before the attack on the World Trade Center. In an emergency, there is pressure on the rescue services to work smart and fast. There is little time for them to assimilate important 3D information regarding buildings, location of combustibles, underground services, overhead landing patterns, etc. Professionals just need to be able to 'see it'. One consultant put it:

> You don't just need the data, you need to visualize this entire thing because you need to send the right crew with the right equipment for their own safety, and for the neighbourhood's safety.

Emergency response management applications have been researched using the VR model of the Los Angeles area (Jepson *et al.*, 1997). It is argued that in the field, the response personnel will be able to accurately locate the position of beacons such as fire alarms, identifying first the building and then the floor and exact location on the floor (e.g., main corridor) visually. The system will then further allow the response personnel to dynamically access databases (maintained by the building owner and occupants), showing plans of the floor plates and the locations of features such as fire hydrants, hose bibs and toxic chemical storage areas. Whilst dynamic movement through VR models may be useful for training emergency personnel, in a real emergency it is not seen as valuable. Remembering sequences of views is more computationally intensive than remembering a single view with all the relevant information. Emergency services need to be able to comprehend all the relevant information, and virtual reality allows rapid and unrestricted movement around a model in order to choose the best vantage point from which to understand a problem. Some researchers believe that use of a real-time 3D model in the process of rescue will revolutionize the way that emergencies are handled by giving emergency services rapid access to information (Jepson *et al.*, 1997).

The base model of New York, mentioned in the section 'CAD and GIS data', and 3D laser scans of the site of the World Trade Center were used extensively in the recovery operation after the September 11 attack. Fire fighters rotated and viewed 3D models of the site to obtain information of inaccessible areas (Sawyer, 2001).

Infrastructure management

Municipalities are interested in using virtual reality to explore the extensive spatial data that they hold and to help them manage their infrastructure. For them the information is of key importance and they are interested in tools that allow them to mine into and explore it. In a discussion regarding how the base map of New York could be extended, one consultant felt that virtual reality was of great use but said that its use had to be tied to a consideration of some of the economic, structural and institutional issues. Without close ties between data and visualization, this consultant felt that visualization was fun, but not useful for change. It was adding links between the visualization and data that made it useful.

Companies that own extensive infrastructure are looking to 3D models to help them manage this. Telecommunications companies, concerned with visualization of their mobile communications networks, have been investing in the creation of 3D models. For example, the Finnish telecom companies Elisa Communications Corporation and Comptel, are part-owners of the city-modelling company Arcus Software.

In New York, development of city models that allow companies to visualize their common infrastructure has been funded by the telecommunications and real estate industries. The company U-Data Solutions sells or licenses its models to companies that need to see and to integrate 3D objects and their associated links to data. The models themselves are quite abstract and symbolic, for example the colour green signifies 'park'. Models are not photo-realistic, so attention is focused onto relevant features.

For management consultants and lawyers the cost of personnel is high and facility managers spend some time looking at access to transportation and amenities before a move. A firm looking to relocate can use these models to assess available space – they can visualize vacant offices in relation to the location of their competitors and in relation to the parking lots, garages, etc. (Plate 23).

Large companies that own and maintain extensive infrastructure are also looking at 3D representations of that infrastructure. For example, Railtrack, a company that owned and operated the rail infrastructure in Britain until 2001 was also in the process of developing a 3D model of the entire network in the National Gauging Project (Glick, 2001).

Planning

Virtual reality is being used in the planning of cities and urban areas. It promises to help us understand the dynamic functioning of the city, just as it has been used to understand other systems, such as factories, airports and shopping malls.

Cities that are great are not those that are just well zoned with the right amenities: the streets of great cities are interesting places to be. The bird's-eye view is blind to the real beauty and power of great cities (Jacobs, 1961; Johnson, 1999). By constraining the user to a viewpoint outside the model, the quality of streets as places to be is ignored in zoning maps as well as in the computer game SimCity (Johnson, 1999). Yet in virtual reality it is possible to take a range of viewpoints within the model, showing realistic views of the streets that make up cities and urban areas.

By focusing attention on the street rather than on the view from above, virtual reality may allow greater participation and better decision making in the planning process. Lynch writes that:

> The metropolitan region is now the functional unit of our environment, and it is desirable that this functional unit

> should be identified and structured by its inhabitants. The new means of communication which allow us to live and work in such a large interdependent region, could also allow us to make our images commensurate with our experiences (1960: 112).

Virtual reality can be seen as providing images to support bottom-up, rather than top-down interventions, as strategies for defending plurality within cities. A supplier said that virtual reality could be used for 'Getting people to buy-in or at least understand the options'. Some municipalities and not-for-profit planning organizations see the use of virtual reality as part of a move to give easy citizen access to information on planning issues.

Leaving planning solely to private interests and dynamics of the market would exacerbate social inequalities, increasing the differential access to communication and transportation networks. Yet infrastructures are increasingly maintained and developed by private companies. Planners and modellers argue that it is good business for the city authorities to make available urban data that can be used in the planning process. They are using information technologies to make their metropolitan areas more attractive to business by streamlining planning processes, removing the risk associated with uncertain and lengthy processes.

Three-dimensional images are powerful communication tools but, like other representations, their use is not completely objective. Model makers may have vested interests and make assumptions in the creation of models. Bosselmann (1999) remarks that anyone preparing and using images in decision making must ensure that the representations are open to scrutiny and independent tests. However, Hall (1993) argues that the use of computer visualization is more objective than the use of perspective drawings as the viewpoint is infinitely variable and can be selected by any party.

In this section we will look at the use of virtual reality for community planning and policy by municipalities and not-for-profit organizations. We will also look at its use by private developers for planning approvals.

Urban models 'Hollywood-style'

Before the term virtual reality was invented, the desire for a participatory planning process motivated some organizations to look at new forms of representation. The Environmental Simulation Center, in New York, USA, was experimenting with Hollywood-style special effects to facilitate planning decisions. They filmed photo-realistic walkthroughs of potential places. Video footage moving through the model at eye level was created using physical models and large-scale machinery for lighting and cameras. Rather than advocating particular solutions, they aimed to work with communities, creating models to facilitate decision making through brokering and consensus building.

As a not-for-profit organization they found these models good for involving communities in planning. However, they faced the problem that the simulation was viewed as an image. It was not easy to ask questions such as 'What would the place look like under different conditions?'. Shadow analysis was not possible as the lighting used was specifically designed for movie making. Change was very costly, and there was a great deal of effort expended on things that did not impact on the planning decisions, such as the ways to take the tops off tall models of buildings so that the gantry would not knock them over. For these reasons, in the early 1990s, the Environmental Simulation Center was looking for more flexible ways of allowing communities to understand planning alternatives.

When staff at the Environmental Simulation Center heard about Bechtel's programme called Walkthrough, which allowed real-time interaction in a 3D environment, they tried to apply this to urban design. In 1992–93, the Center underwent a transition from using gantry-based photo-realism and cardboard as the means to achieve their ends, to using high-end computer visualization.

5.1
The gantry-based simulation system used at the Environmental Simulation Center

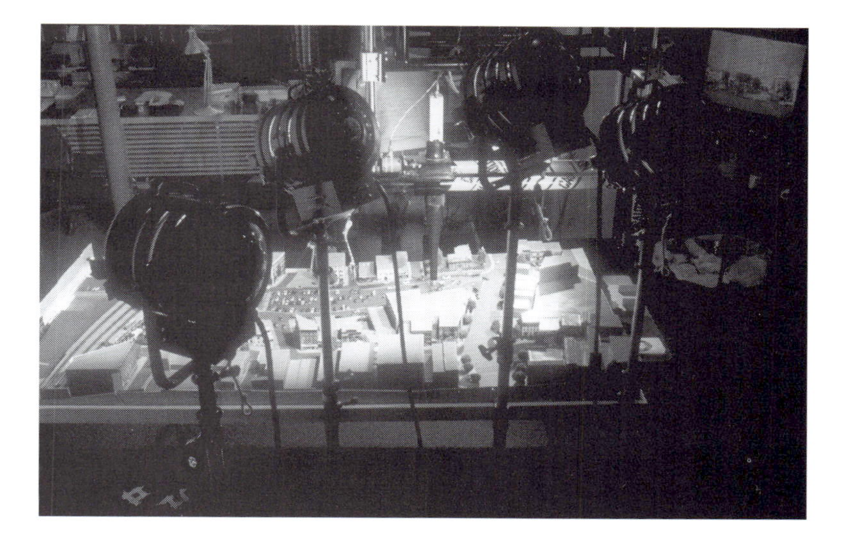

5.2
Lighting a model for video presentation, using the gantry-based simulation system at the Environmental Simulation Center

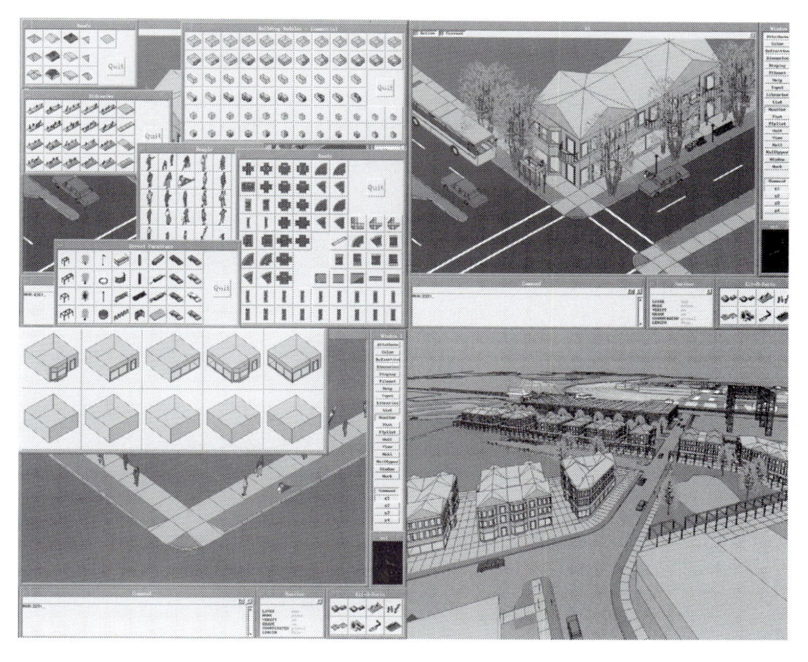

5.3
Kit-of-parts used on the Princeton Junction project by the Environmental Simulation Center

One of the early projects done using the computer was the visualization of a new development for Princeton Junction, an area between New York City and Philadelphia on a commuter and Amtrak line. The aim was to combine words, numbers and images, so that visuals were backed by information about the places that they represented. This project took a kit-of-parts approach and developed some of the early concepts associated with 3D GIS. The models allowed different groups to query and reuse the model. For example the transit authority might want to query the model to ask: what population do you need to use this transit system? Which transit stops do you want to keep open at night?

Community planning and policy

Geographic information systems and virtual reality are being used to increase the input of residents and interested parties into the planning process and to enable a greater range of alternatives to be evaluated. They make it easier to ask questions about future developments. Companies such as the Environmental Simulation Center and IT Spatial aim to use impact analysis in a real-time environment to allow policy forecasting. Automated impact analysis tools can be used to query the environment, asking questions about the number of school-age children within an area, or the number of trips to the nearest metro station, etc.

Typically, planners look at urban massing using wooden blocks, isometrics and sketching tools. At present planning policy is decided by using a 2D layout, then looking at the implications, and finally asking what that looks like. In the USA, the planning is constrained by simple factors, such as height and density. The Environmental Simulation Center believes that instead of having public policy first and then designing the place, it is possible to start with what the place looks like and then work back. The visualization tools that they develop can be used to understand the implications of policy.

Interactive, spatial, real-time tools allow planners to look at urban massing in the context of the existing developments. In VR applications that are dynamically linked to GIS, clicking on a building or object in the interactive 3D view allows the user to access any data related to that building or object that is stored in the GIS application.

Staff at the Environmental Simulation Center have developed techniques for getting communities to consider different development options. They see virtual reality as one tool within their toolkit of methods for sharing alternatives and enabling people to interact with planning options. Representations in virtual reality are used alongside other representations, such as animations, photographs and CAD drawings, to improve understanding of space. For the Environmental Simulation Center, the important thing is enabling communities to understand the 'experiential' aspects of place (Kwartler, 1998).

A key lesson is that virtual reality cannot be used in an unstructured manner but must be part of a larger narrative about the potential options. It can be used to facilitate a discussion in a structured manner in conjunction with a range of other tools for involving all participants. The staff at the Environmental Simulation Center first try to understand the audience and what interactive, real-time, spatial media are being used to achieve. They then use the models to ask questions. The use of models depends on the context; whether the models are used in a series of workshops or a

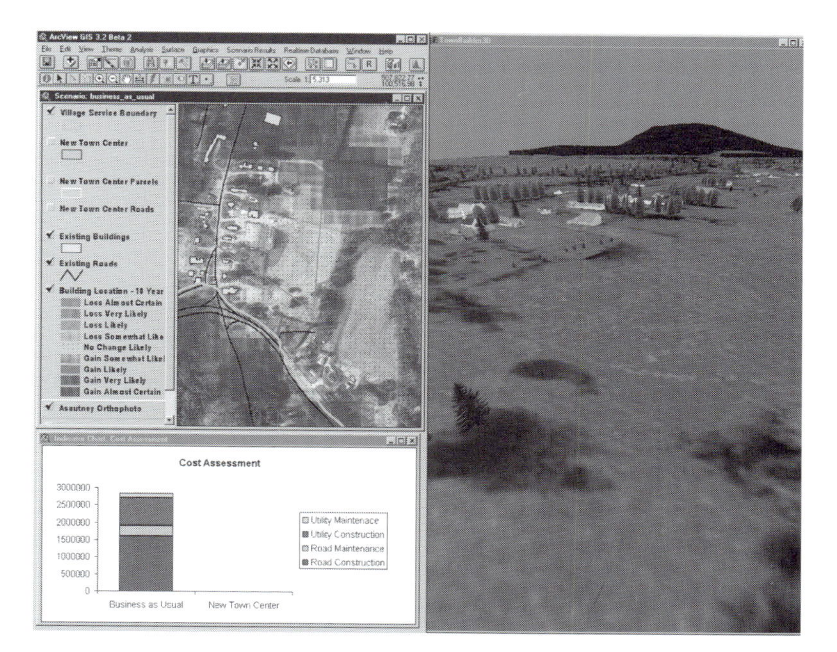

5.4
Community Viz showing a landscape before development

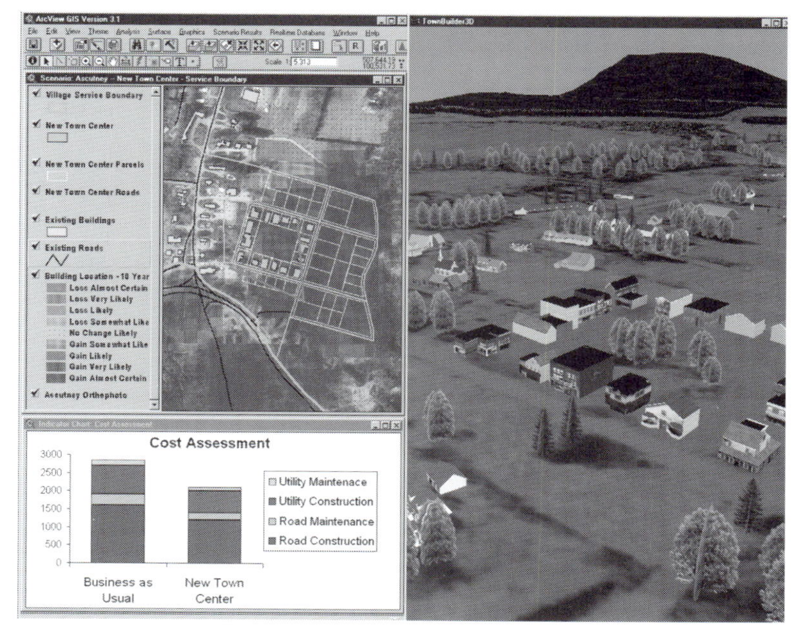

5.5
The same landscape after development, as shown in Community Viz

one-off meeting. The chief planner at the Environmental Simulation Center points out that real places are far more intriguing and complex than anything that can be created on the computer. After considering the simulations this planner often says to people that they should step outside and consider what the real place could be like.

5.2 Santa Fe, USA

The Environmental Simulation Center collaborated with planners to develop alternative growth scenarios for an area of Santa Fe in New Mexico. This is an area that has been developing with a scale and character at odds with the rest of the community. The Environmental Simulation Center designed and modelled new street and neighbourhood patterns as a series of building blocks that met community goals and could be easily incorporated into the community's existing development patterns.

The model that they created was used to show alternative densities in meetings with local residents. One of the densities was seen as too high but, after using the simulation, the local residents selected a density that they would have otherwise never have thought of agreeing to. The use of virtual reality allowed them to visualize the implications of density and to come to a new understanding of the nature of place.

The integration of GIS analysis, urban design principles and 3D visualizations within the public participation process enabled consensus on planning and design principles. Residents had been given more information, which meant they were able to consider the wider implications of their decisions. The final solution of higher density neighbourhood centres mitigates the detrimental impacts of sprawl while meeting projected housing requirements.

5.6
Comparing different options for Santa Fe – buildings at the front or parking at the front

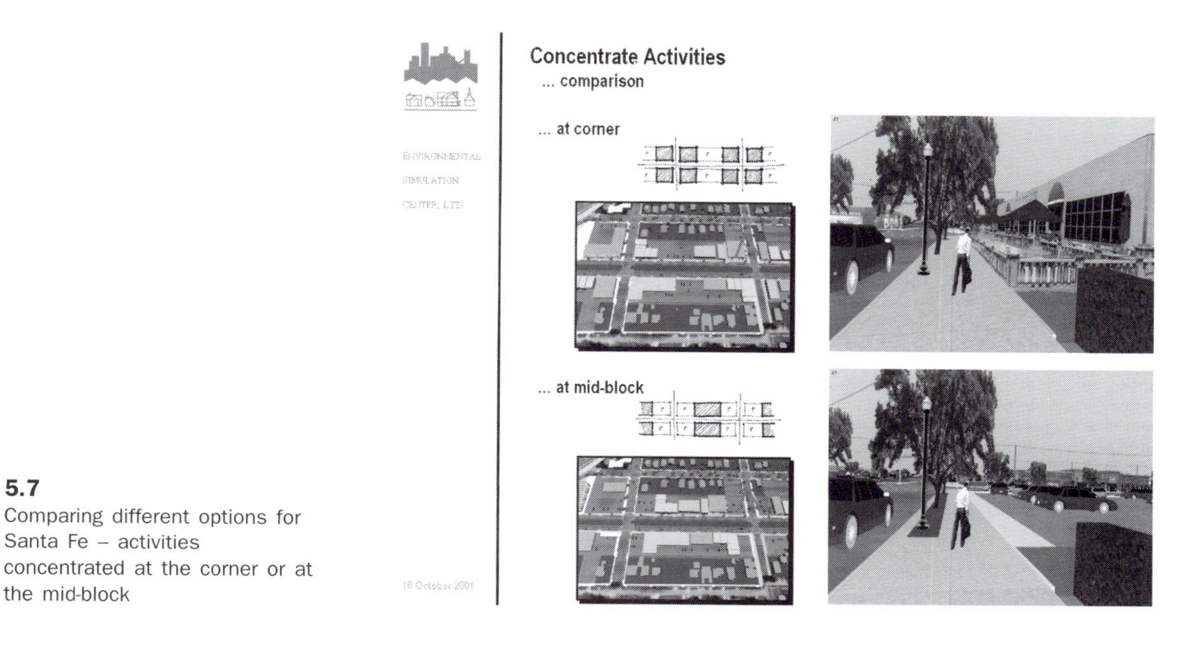

5.7
Comparing different options for Santa Fe – activities concentrated at the corner or at the mid-block

Regeneration and location marketing
Marketing is a driver for the use of virtual reality in both municipal authorities and industrial organizations. Virtual reality is being used for regenerating areas and for marketing locations.

Virtual reality has been used for regeneration in Chemnitz, Germany (Figures 5.8 and 5.9). In the era after the fall of the Berlin wall, this mid-sized town faced a number of problems with the management of its urban environment. People who could afford to move out of the old socialist housing estates were doing so. During a period of renovation, when the area might be unpleasant to live in, planning officers felt that they needed to persuade people to stay in the area. They wanted to be able to show people what the area would be like to live in after it had been renovated and they commissioned a VR model of the city from the company Artemedia. This virtual model has been used to facilitate discussion about how the city centre and socialist housing of the communist era can be changed and upgraded for modern living.

The Department of Planning and Permitting at the City and County of Honolulu in Hawaii, USA, is continually challenged to interpret proposed plans for developments and buildings. They have to assess their potential visual impact on adjacent properties and surrounding views. The

5.8
The historic centre of Chemnitz, showing locations of proposed new developments in a VR model developed by Artemedia

5.9
A view of the VR model of Chemnitz, showing transportation routes through the centre

proposed redesign of the Kapiolani Boulevard and Kalakaua Avenue intersection in Honolulu was a particularly important project. Though the area had a lot of old buildings and was run down, the intersection formed a gateway to all of Hawaii as most tourists passed through it on their journey from the airport.

The Department was able to build a VR model of the Kapiolani Boulevard and Kalakau Avenue intersection gateway in about two and a half weeks and this model impressed City and County of Honolulu officials, helping the Department to secure the necessary funds for the project. The model was also displayed at the Mayors' Asia Pacific Environmental Summit 2001. Because of the success of the project officials are considering a 'virtual permitting' programme for Honolulu, requiring each devel-

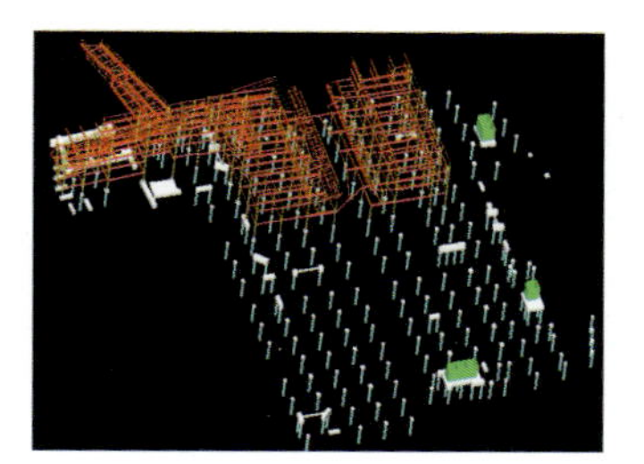

Plate 12
The model of Basingstoke Festival Place, created by Laing Construction, with selective loading of subsystems

Plate 13
View of Farringdon station, London, UK in the Bechtel model of Thameslink 2000

Plate 14
View of a railway cutting from the driver's viewpoint in the Bechtel model of Thameslink 2000

Plate 15
View of a signal at red from the driver's viewpoint in the Thameslink 2000 model

Plate 16
View of the same signal at green from the driver's viewpoint

Plate 17
Times Square in New York showing information displayed on adjacent buildings

Plate 18
Interior view of the trans-ports project by Oosterhuis

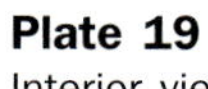

Plate 19
Interior view of Oosterhuis trans-ports project showing people interacting with and generating dynamic images

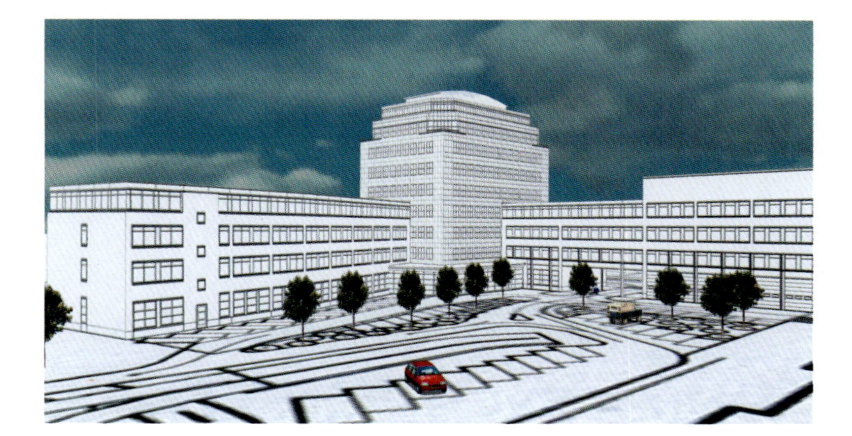

Plate 20
A VR model of the Westeinde hospital design by Prent Landman

Plate 21
Another view of the Westeinde hospital, Holland in the VR model

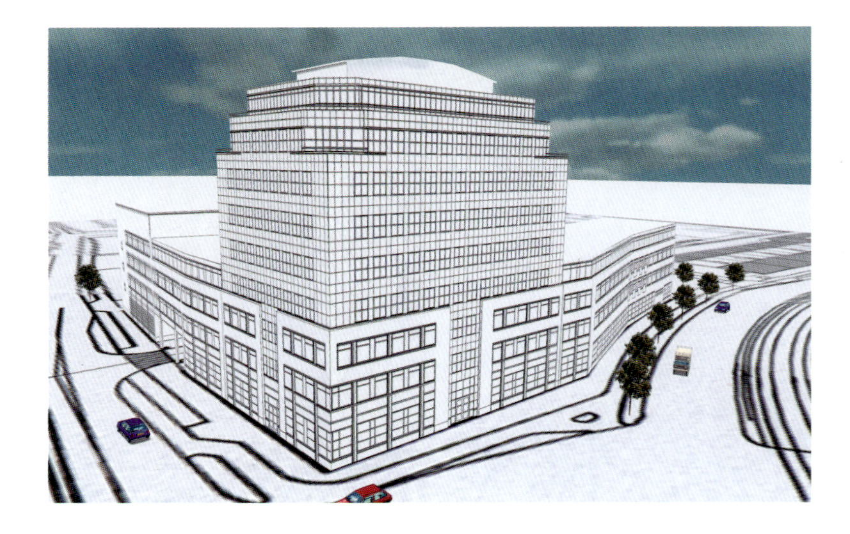

Plate 22
The model of Los Angeles, USA is developed using the MultiGen-Paradigm modelling tool and will eventually cover the whole basin, an area comprising more than 4000 m^2

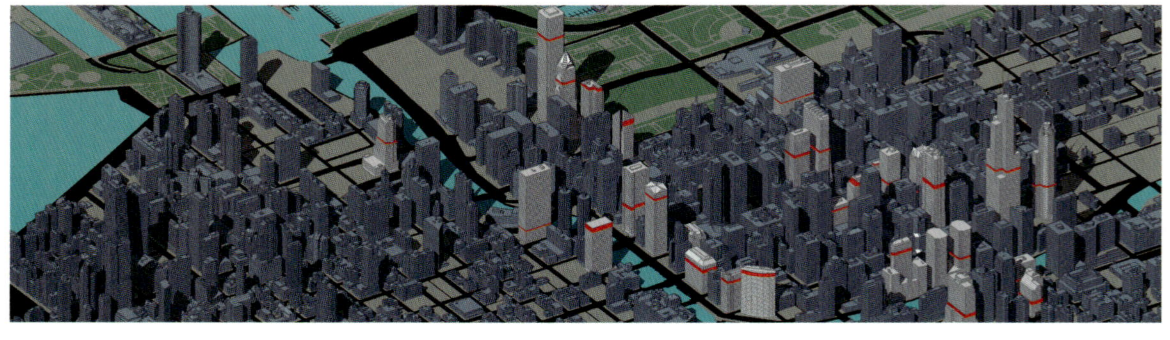

Plate 23
Model of Chicago, USA, by U-data solutions, showing the location of major law firms

Plate 24
Virtual reality model showing the first phase of the planned community, to be built near Salt Lake City, Utah

Plate 25
Virtual reality model showing the first phase of a planned community, to be built near Salt Lake City, Utah

Plate 26
The model of Berlin, Germany. On the VR model, colour is used to differentiate between built and not-built projects. On given lots in Berlin the regulations indicate that skyscrapers will be allowed and this is shown on the model

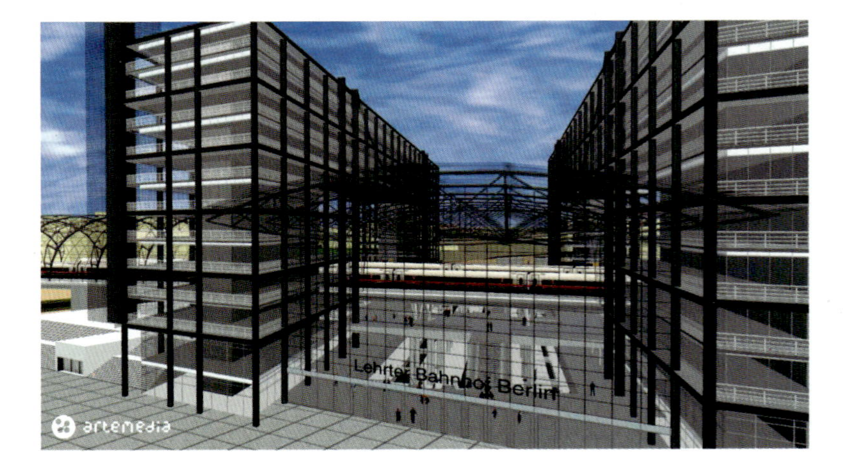

Plate 27
The model of Berlin, Germany showing the new train station, by von Gerkan, Marg and Partners

Plate 28
A massing model, representing Charlotte, North Carolina. The model was created by Skyscraper Digital and shows the geometry of environment

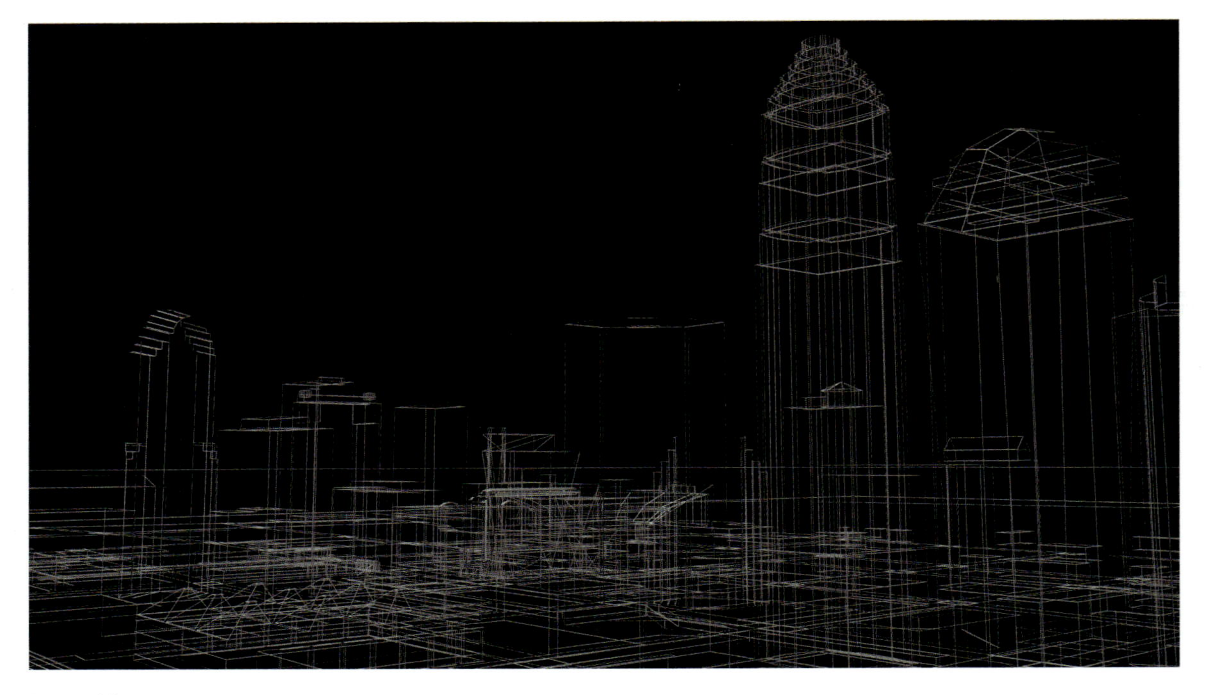

Plate 29
Skyscraper Digital's 'Charlotte City Model' as a wireframe model

Plate 30
A fully rendered and texture mapped version of the 'Charlotte City Model' showing the virtual environment after it has been rendered and digitally painted

5.10

5.11

5.10 and 5.11
Images from the virtual model of Honolulu

oper or builder to submit 3D source data for their proposed project.

Virtual reality has also been used to market new locations. Kodan, the Japanese government developers, developed the new town of Kizu, which is located near Nara in the Kyoto prefecture in Japan, and used a computer visualization to allow many people to view the scheme and freely navigate through the model. To aid navigation, the interface to the model included a north-up map, which showed the field of view. As the model was designed to allow many users, the viewpoint was reset after 300 seconds, to allow the next person to use the model. This model was used at the opening ceremony and placed at the station to the new town to allow for public viewing and orientation.

5.12
Bird's-eye view of the new town of Kizu, Japan

5.13
View of a street within the new town of Kizu, Japan

5.3 Harvest Hills, USA

When a land developer decided to develop a master-planned community covering 1.42 ha on the north shore of Utah Lake, near Salt Lake City, consultants were hired to build a model of the development to present to the planning commission and city council. The development, worth US$150 million, was being developed according to a tight schedule so the developer was keen to obtain approvals quickly. Being able to walk the planning commission and the city council through the development speeded up the process. The developer said:

> *Presenting plans for approval is always very stressful for us. We are always concerned that the commission won't be able to visualize the development from our drawings and sketches.*

The use of virtual reality allowed this developer to reduce the delays and related business risk that submitting planning applications often involve (Plates 24 and 25).

Planning approvals

In the early 1990s Chelmsford Borough Council in the UK recognized the potential of computer visualization to facilitate the work of planning authorities by removing barriers to communication and thereby assisting negotiations (Hall, 1993). At the end of 1991 they had received a planning application for a detached house in Danbury and one of the applicants, her agent and a planning officer assembled around a computer monitor displaying a very basic model of the house and neighbourhood properties. All parties found this useful and the planning officer commented that it enabled people who had not visited the site to participate fully in the discussion.

Delays are expensive for developers, and in many countries the length and uncertainty of the planning approval process adds considerably to developers' business risk. In the UK some developers are interested in using virtual reality to reduce delays in the planning process (Whyte, 2000). However, planning authorities vary in different regions of the country and some UK-based housing developers felt that the planners in their region looked unfavourably on computer-drawn images and models. One CAD manager thought that the planning authorities liked the technology but did not want to be challenged to use it:

> I truly believe that some of the local authorities are a little bit scared of the technology when it comes to operating it ... but they like to see it.

Another national house-building company used a visualization package to create a hand-drawn look on drawings submitted for planning approval, explaining that 'planners criticized the CAD drawings as too regimented when [we] started using the computer for design' (reported in Whyte, 2000). It may be that these developers can learn from the more sophisticated users of virtual reality in Japan. To include novice users, Japanese house-builders use a range of representations.

There is an opinion that virtual reality and other computer-based visualization techniques must be used with care with the local planning authorities. With virtual reality there is the potential to see views that can not be seen in real life. A housing developer produced an animation to show that the 3D changes they had made to the site meant that these views for other residents in the area were not altered. The developer noted that if the planner were able to pick an unrealistic eye-level this could have caused them considerable delay in proving this point.

Housing developers were also worried that giving more information raised the possibility for further questioning. They felt it might shift the planners' attention away from the relevant planning issues and 'they will ask questions about materials, particular stonework, etc., that they would not ask if you didn't give them the information' (reported in Whyte, 2000).

A greater ability to conceive and imagine design can lead to a greater ability to misrepresent that design in the adversarial arena of planning approvals (Bosselmann, 1999). To use the model on a legal process, which development assessment is, requires that the model is credible and verifiable (Pietsch, 2000).

Drivers, barriers and issues

Technological trends are increasing the amount of data available for the management and planning of the urban environment. The use of these models may improve the quality of the built environment, but this use raises many issues, including those related to misrepresentation, intellectual property rights, data security, appropriate financial

5.4 City model of Berlin, Germany

In the German capital, Berlin, a PC-based city model has been commissioned by the authorities and created by the company Artemedia. The basic model was created from land registry data, containing information about the city lots and services such as gas and water. Aerial photographs were superimposed to gain height data. The resultant model shows the location of street blocks and key buildings rather than detailed models of all buildings.

This model of Berlin has been used in the planning and development of a number of high-profile projects. Developers are very interested in including their new schemes in the model and they will pay to have models of their development built and included. Von Gerkan, Marg and Partners used the model of Berlin when designing the new train station, Lehrter Bahnhof, in Berlin. Foster and Partners also used it when they were working on the Reichstag building. It was used in the creation of Potsdamer Platz, and the marketing of the spaces it contained before the construction was completed (Plates 26 and 27).

However, the model is intended to represent the current state of the city so new developments are not introduced unless they have the approval of the city. Artemedia manages the model for the municipality and they ensure that nothing is introduced unless it has been signed off by the city.

models and ownership. At the urban scale, there are drivers for the use of virtual reality in both the management of urban infrastructure and the planning of new developments. Whilst people find virtual reality useful at the urban scale there are fewer short-term financial benefits to the organizations that commission the models. Budgets are lower, but advocates believe there is the potential to increase the quality of the design and management of the built environment. Major issues include ownership and access to models and the extent to which the use of the models enables participation in the planning process.

For the creation and effective use of urban models for public good, the use of intellectual property rights is an issue that needs to be dealt with. The way that mapping agencies and other bodies that provide the data charge people for that data changes they way in which data are used. Companies can be loath to share their information, and in many cities and urban areas there is no common base map onto which all city data can be superimposed. Many attempts to integrate sources of information about urban environments are motivated by a desire to make government more transparent, increasing the information available to citizens. However, for reasons of security and commercial sensitivity, some data cannot be made public. There are issues regarding the extent to which virtual reality allows citizens to visualize alternatives.

The development of some city models has been privately financed. Skyscraper Digital has developed a city model of Charlotte, USA, through work with a number of private companies. The two major banks in the city of Charlotte, the Bank of America and First Union Bank, as well as other companies such as Cousins Properties and Discovery Place, have funded the development of city blocks. Skyscraper Digital looks to retain ownership and copyright of the city model, licensing it to the clients rather than giving them the source models.

The business model they used to create this was to work for clients on projects for particular buildings within the model and then use the budget from those projects to finance the further development of the surrounding areas of the model. This type of business model has resulted in a virtual city in which some areas are very sparse whilst others are very detailed, as these are where the clients have been focusing their projects.

Ownership of city models is a key issue. Academic projects have shown the potential of such models at the urban scale and provide good case study examples, but few models built in academia are being used to their full potential in the planning process. This is partly because the city authorities and other relevant parties do not feel that they 'own' them, as they have usually not been involved in their development (Watson, 2000). Municipal authorities are beginning to use virtual reality in-house, or are working in collaboration with suppliers to develop and maintain city models.

The growing interest in collecting and using 3D data at the urban scale, discussed in this chapter, will have practical implications for the organizations involved in the design, production and management of the built environment. It is to these that we turn in the last chapter.

6 Practical implications

Many of the organizations involved in the design, production and management of the built environment are project-based (Gann and Salter, 2000). They are involved in a portfolio of projects, using their skills and expertise on each project over its finite lifetime. In this final chapter we look at the use of virtual reality within these organizations and explore practical implications of its use.

We have seen how the Schools of Cartography in the Borges (1946) story became interested in the Map the size of the Empire, without regard for its application. In contrast, the professionals interviewed in this book are not interested in virtual reality itself. Instead, they are interested in what they can do with it. As the case studies show, some leading organizations are using virtual reality to differentiate their products and services within the sector. Others are using it to diversify their interests and exploit new market opportunities for spatial design skills.

Virtual reality provides professionals with an interactive, spatial, real-time medium. Its use may lead to a convergence between packages for product prototyping and those for process simulation. Yet virtual reality is not being used as a generic technology in leading organizations.

Companies have found practical applications for the use of virtual reality across a range of different activities: demonstrating technical competence, design review, simulating dynamic operation, co-ordinating detail design, scheduling construction and marketing. However, models created for use within the professional project team and supply chain are markedly different from models created for wider interactions with client, funding institutions, planners and end-users.

- Virtual reality is being used in an abstract and symbolic manner within the project team, to explore the engineering systems and to communicate design. A viewing perspective from outside the model is often used and the interface may include design aids and allow free viewing of the model.
- Virtual reality is being used in a more realistic and iconic manner for explaining design to other parties. Things in these models are made to look like the things that they represent. The models show surface detail and are often used to explore aesthetic considerations, such as external appearance, interior decoration and furnishings. A viewing perspective of a person within the model is often used and interaction may be guided and supported by predetermined viewpoints, etc.

Table 6.1
Attributes emphasized in models for use by the professional project team and models for wider involvement

Professional project team	*Wider involvement*
Abstract	(Photo) realistic
Symbolic	Iconic
Engineering system	Surfaces
Exocentric viewing perspective	Egocentric viewing perspective
Rapid design changes in real-time	Fine-tuning of design offline
Design aids	Navigation aids
Free-viewing	Controlled viewing

Table 6.1 shows the attributes emphasized in each case. These findings raise questions about the extent to which different functions will become integrated. Generic VR applications may not be the only mechanism by which interactive, spatial, real-time techniques diffuse through the construction sector. Interactive, spatial, real-time views may become used in a wide range of professional applications, and the use of the VR medium may lead to more intuitive interfaces to data.

In this chapter we will look at how the project-based nature of production affects the ability of companies to implement and use virtual reality. First, the nature of design visualization in the project-based firm is explored. Then the industrial context and issues raised in the book are considered and the reorganization of practice is discussed. Finally, a checklist of factors that affect the application of virtual reality is given. This checklist may act as a starting point. It is expected that it will be extended and refined as we learn more about applications of virtual reality in the design, production and management of the built environment.

Design visualization in the project-based firm

Design practices are not homogenous across the construction sector. As described in Chapters 3, 4 and 5, many different types of organization are beginning to explore the use of virtual reality on a wide range of projects. The business benefits obtainable from virtual reality are affected by characteristics of the projects on which they are used and the design processes employed. Two aspects that may influence organizational choices regarding modelling and visualization tools are:

1 *the size and complexity of the project* – implementations and uses of virtual reality on projects of different sizes,

Table 6.2
Matrix to show the extent to which components of the design are reused and the complexity of the project. Virtual reality is being more widely used on simple projects with design reuse and on individual large complex projects

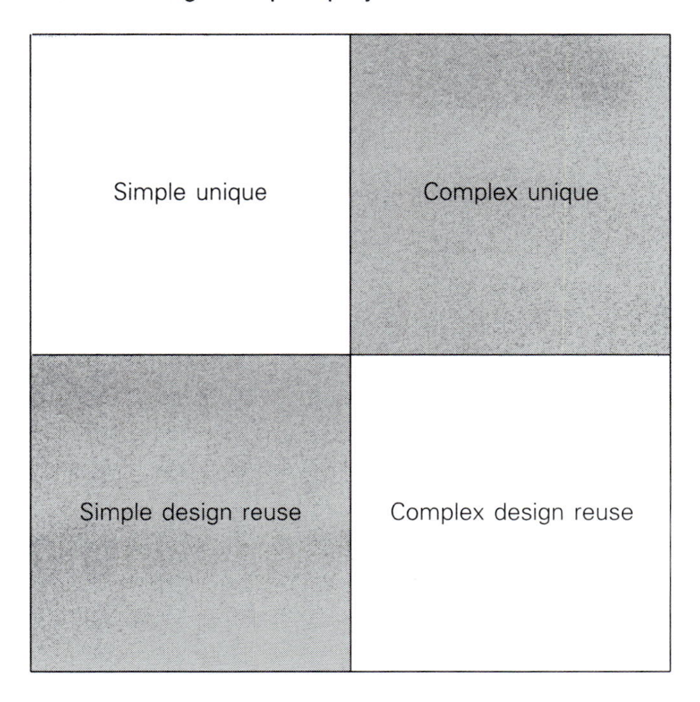

Simple unique	Complex unique
Simple design reuse	Complex design reuse

6.1
VR model of Dubai International Airport by Bechtel

6.2
Another view of the VR model of Dubai International Airport

relative to the size and capacity of the project-based firm, present their own particular problems; and

2 *the extent to which components of the design are reused* – some projects have a high degree of standardization, allowing design effort to be reused across different projects, whilst other projects are for bespoke products.

As we discovered in previous chapters, there appears to be more potential for the effective use of virtual reality on large complex projects and small projects in which design is reused.

Individual large complex projects

Companies that work on large complex projects have major business drivers for the virtual reality within the project team and supply chain. On these large complex projects, professionals, such as consultant engineers and construction managers, use virtual reality to visualize and understand engineering problems and hence to reduce risk and uncertainty.

Large complex projects may involve the collaboration of a number of different specialists and organizations over an extended project lifetime. During this period, teams can act as a single organization through collocation, secondments and close trust-based working relationships (Gann and Salter, 2000). Budgets for hardware and software may be relatively large and there may be a greater investment of time in model building. Modelling and visualization staff may be seconded to work on the project full time and models may become a focus for design and a repository of design knowledge. These models may be returned to over an extended period and used for integration of different sub-systems and design checking.

Small projects with design reuse

Companies that work on small projects have gained benefit from using virtual reality at the customer interface when they have been able to reduce the resource input by reusing models on many projects.

In these companies, the modelling, visualization and design staff may be working simultaneously on many projects that are in different stages of development. The CAD managers in a major house-building company, for example, typically have about thirty projects on their desks at any point in time (Whyte, 2000). They return to work on individual projects periodically as they move through the design, planning and construction processes.

Virtual reality has been used successfully on such small projects, though budgets for hardware and software are low and few hours can be invested in model building on individual projects. Low-end VR and interactive 3D have been used successfully in organizations where some design standardization allows components to be reused. A successful example is the use of interactive 3D for the marketing and sales of housing in Japan.

Industrial context and issues

As discussed, leading organizations are gaining benefits from the use of virtual reality on both large complex projects and small simple projects with design reuse. On these projects they are able to reuse models to reduce the resource input.

Yet the industrial context within which organizations operate is changing as suppliers begin to develop and customise VR applications; competitors begin to use virtual reality; and planners, regulators and customers begin to demand or expect its use. Changes in this wider context may lead to the technology being used across a wider range of projects. Factors, such as the procurement route and the risk of a project not reaching completion, may also affect the use of modelling and visualization tools. These and the changing context within which project-based firms work affect the incentives for organizations to use virtual reality and the benefits that can be obtained.

Suppliers, regulators and customers
The emerging market for interactive, spatial, real-time software is young and dynamic. Suppliers of high-end VR software are looking at construction as a potential growth sector and are competing in this market with both new entrants and the CAD suppliers that are incorporating interactive, spatial, real-time characteristics into their packages. As yet there is no dominant business model and companies are switching between licensing their models, selling model building services and selling their software packages.

Project-based organizations that have competitors using an interactive, spatial, real-time medium may revisit their own corporate strategies. As we saw in Chapter 5, one influence on corporate strategy may be the urban planners and managers who are describing virtual reality as infrastruc-

6.3
The virtual model of the City of Jyväskylä is generated automatically from digital maps using NovaPOINT Virtual Map software. The system is used in city planning and decision making

6.4
A wide screen can be used to present virtual models to bigger audiences, giving them a different perception of the scale and perspective that is important in environmental analysis and city planning. This image is from Aalborg VR Media Lab

ture for decision making. Planners and regulators may demand more information from the project-based organizations involved in the design, production and management of the built environment as they benefit from uses of virtual reality at the urban scale.

Clients may also become more demanding. Virtual reality promises to offer clients, managers and end-users greater ability to input into the design process. By enabling users and designers to discuss design issues, virtual reality may be used to enhance the quality of the final product. We have seen that, for companies that own, design, build and operate real estate, investment in tools such as virtual reality is seen as a commercial decision to spend in capital in order to save in running costs.

Wider uses

The changing industrial context may lead to wider uses of virtual reality within the construction sector and in new emerging markets for spatial skills. Whilst leading users of virtual reality have worked on large complex projects and small projects with design reuse, virtual reality may become more widely used as designers find new ways of gaining benefits and reducing resource inputs.

In the short term, the risk of a project not reaching completion is a factor that may affect the use of virtual reality within organizations. At the early stages, design work is inherently full of uncertainty and risk. Many of the projects at the viability and simple design stages may not progress to full completion of the final building. For architectural practices, this dropout rate can be as high as 50 per cent. Projects also have long lead-times. Those that reach the detailed design stage normally go forward to full completion. However, this may be a lengthy process as timescales are dependent on business cycles, as well as on planning processes. The risks involved with the dropout rates and long lead times affect the investment in and use of visualization within organizations. There is a need for these organizations to get buy-in from their clients during extended building schedules, and to get their money back on projects.

The risks of projects not reaching completion compound the problems of using virtual reality. However, despite these risks and the unique designs that they produce, some architects are starting to use modelling and visualization tools. An example of a lead user of 3D is Kohn

Pedersen Fox, which has created a model of London that it is being reused across a number of projects. This type of reuse may show how early users of virtual reality can obtain business benefit. As planners, regulators and clients demand more information, designers will find new incentives for the use of virtual reality.

6.1 Kohn Pedersen Fox (KPF) model of London, UK

In some circumstances, investment of time and money in large 3D models can be spread over a range of projects that are one-off designs. As a large commercial architectural practice, KPF do a lot of work in the city of London, designing large schemes such as the AIG headquarters and Heron House developments. They have a large and comprehensive 3D model of the city. It has been created in the Microstation CAD package and serves as a valuable resource for the practice with considerable reuse value. At any point in time KPF are simultaneously working on different projects within the model of the city and architects within KPF can use the model as a resource.

As well as potential business benefits in the design, production and management of the built environment, interactive, spatial, real-time media open up new market opportunities to architectural practices. Conceiving of these media as 'electronic space' is useful commercially, allowing the architects to market their expertise in spatial design. Virtual space allows them to explore spatial concepts without concern for the messy happenstance of lived-in reality and without the clients, end-users, fabricators and schedules that are associated with physical buildings. This brings architects (reconceived as cyberspace architects or architects of the physical and virtual realm) into competition with Web-designers, human–computer interaction experts and programmers and further removes them from their original concern with the design of inhabited places and the social, economic and political implications of the built environment.

Issues

The use of virtual reality for the design, production and management of the built environment raises a number of cognitive, technical and organizations issues.

In Chapter 2 we discussed some of the cognitive issues by looking at how virtual reality is different from reality, and considering how representations in the virtual reality medium can be useful in problem solving. The different types of exocentric and egocentric viewing perspectives, modes of navigation and performance aids that can be used in virtual reality were described. In later chapters we have seen how leading users are learning to use virtual reality in an increasingly sophisticated manner across a wide variety of tasks. Effective use of virtual reality is seen as task-dependent and learning and experience are seen as important factors.

Technical issues such as data translation and the strategies for model creation – library, database or straight translation – were first discussed in Chapter 1. Construction sector users have not influenced the early development of the VR system. However, the emerging business drivers for the use of virtual reality in the design, production and management of the built environment may start to shape technological development of virtual reality as software suppliers gain customers within the sector. Users are a good source of innovations (von Hippel, 1988) and innovation often occurs at the boundaries between different traditional roles (Hobday, 1996).

Organizational issues are highlighted in many of the case studies with leading users of virtual reality. Virtual reality can be used to informate processes within the organization, but successful models may become enshrined in particular departments instead of being shared more widely. In the next section we will look more closely at the use of virtual reality in organizations and ways in which these problems may be overcome.

By looking at the emerging uses of virtual reality in industrial practice, rather than its use in the research laboratory, this book has tried to shed light on some of the cognitive, technological and organizational issues that its use raises. As the use of virtual reality becomes more established we will learn more about these.

Other issues that arise result from the fact that new technologies have dual uses. For example, whilst virtual reality may be used to increase participation, it may also be used to reduce it. Critics of virtual reality have argued that the idea of the virtual city is about establishing order and coherence – this time in a substitute, or proxy, electronic space (Robins, 1999). Some elements of real life are left out of virtual worlds, so there is an extent to which the models show us only what we want to see. Decisions have to be taken as to what will be included in the simulation and what will be left out.

People ask why there are no beggars in the virtual model of Los Angeles, however there are no business drivers for including them. Thus virtual reality can be utopian, whilst lived-in realities contain the unplanned: beggars and litter and 'the uncertain traces left by events' (Lefebvre, 1974). In a simulation, unwanted aspects are not recreated and unnoticed aspects of an environment cannot be recreated. Robins points out that:

> These are technologies which – in new ways, and perhaps to an unprecedented degree – afford detachment and insulation from the contamination of reality (1999: 51).

We can see that virtual reality does not necessarily increase participation, but it may be used to help participatory design as one tool within a wider tool kit. In this book we have argued that virtual reality is of most use as a prototype that can be discussed, challenged and recreated. For professionals working in this sector, the use of virtual reality as image may be useful for marketing, but there is a danger that image may be used in the design process to seduce rather than to question. As an image virtual reality may become used as an alternative reality and disconnected from any underlying data. As a prototype, the visualization is only useful if it enables the user to understand the underlying design data.

Reorganizing practice

There are many different approaches to the implementation of virtual reality in leading organizations. For some organizations, large complex projects provide the opportunity to focus resources, spend on high technology and pilot new systems. In other companies, the risks associated with new technology uptake are too large to be borne by a single project and may be spread over a number of smaller projects. In the projects studied in this book, three main scenarios for model creation were found:

1 *central technical department* – in some of the companies, a specialist visualization group within the organization championed the use of virtual reality on all projects. This group had often been spun out of the

research and development (R&D) or CAD department and had established a separate identity. The introduction of innovation by central technical departments allows a more strategic approach to innovation and its implementation, but technical staff may fail to get the new technology taken up by those working at the project level;
2. *project based* – in other companies, virtual reality was introduced at the project level by the staff working on a particular project. However, any innovation implemented at the project level may not become known within the organization and be reusable on other projects;
3. *outsourced* – in other cases creation of a VR model was commissioned from, or created in collaboration with, a service provider, such as a university or a commercial retailer, that provided the models and sometimes the viewing facilities to the company. Outsourcing allows companies to reduce their risks, whilst leveraging benefits from the technologies. It increases flexibility, enabling a company to move to different solutions faster as it is not locked into particular sets of technologies because of staff competencies.

Within most companies that use virtual reality in-house, there are a few individuals who act as visualization specialists and model creation is not seen as a generic skill to be learnt by all staff. These visualization specialists may have different competencies and backgrounds to other professionals within the organization, particularly when models are used to communicate with non-professionals. There is a danger that the use of virtual reality does not diffuse across the organization. Consequently there is a need to reward collaborative use of VR models across functional units within the organization so that an innovation ghetto is not created in the sub-unit that has access to them.

In US companies, this type of specialist visualization group is often branded and marketed separately, making it easier for it to bid for external work. For example, Skyscraper Digital, the visualization group within the architects Little & Associates, works on projects for other companies as well as those for its parent architectural company (Plates 28–30).

Outsourcing of technology has often been wrongly equated with loss of skills. Companies' technological competencies are dispersed over a wider range of sectors than production activities and this range is increasing (Granstrand *et al.*, 1997). Skills are required to collaborate effectively with external model-building organizations and to use the models that they produce within the design, production and management processes. In their staff they foster the ability to act as brokers of different solutions, rather than the technological skills to develop solutions themselves. Whilst model production is not a core skill of many of the companies involved, firms need to foster some internal understanding of the underlying technologies in order to use the models well and to explore and exploit new opportunities in their use.

Concluding remarks

The environment within which project-based organizations operate is changing. At the same time, virtual reality is an emerging technology and as yet there is not one dominant design. The established divisions between different types of software are also becoming harder to maintain as technologies bleed into one another.

The practical experience of lead users is the best guide we have to understanding implementation of virtual reality. Industrial use is not widespread and we still have much to learn about the use of virtual reality within organizations. One interviewee said:

> As yet there isn't a set of rules or guidelines ... and I would imagine that as we go along we will start to develop a much more solid series of guidelines for what works and what doesn't.

By looking at the successful applications described in previous chapters, we can develop a checklist of the factors described in this book that lead users have found useful in their implementation and use of virtual reality. These include the following.

1 *Using virtual reality as a prototype* – relating the visualization to the data and using it to develop and test alternative solutions. As a prototype, a representation in virtual reality is a powerful tool for exploring data, rather than an image or alternative reality. It is as a prototype that they are most useful in the design, production and management of the built environment. In this way they enable new product characteristics to be verified and the processes of their construction and operation to be simulated.

2 *Tailoring the model to the task and the user* – considering the appropriate balance of abstraction and realism, the best viewing perspective and additional information. By tailoring the model to the task and the users, organizations can leverage greater advantage from their use of virtual reality. It is possible to explore models from both egocentric and exocentric viewing perspectives and through both abstract and photo-realistic representations of the same data. Some tasks may require the problem domain to be understood in different ways and, as the medium is flexible, the user can move between different views of a model to facilitate their thinking. Novice and expert users vary in their ability to use different forms of representation and novices will require more support when using virtual reality. Navigation in virtual reality can be aided by making landmarks, route and survey knowledge available.
3 *Seeing virtual reality as one tool within a tool kit* – considering how it is integrated with other tools and techniques used. Virtual reality should be seen as one tool within a tool kit. It is one of a range of techniques that can be used to explore options and show a range of alternative scenarios. Lead users are using it in conjunction with other forms of representation to see problems in different ways and to involve all in discussion of design.
4 *Developing structured ways of using virtual reality* – using it to support organizational aims and narratives about design. The medium is at its most powerful when used in a structured manner and successful early users have used it within a structured process that is focused on achieving particular organizational aims.
5 *Looking for opportunities to reuse modelling effort* – taking into account the nature of the projects that the organization works on and the risks associated with dropout rates. Organizations can gain by reusing modelling effort on individual projects, such as large projects where there are complex products; and across many projects, such as small projects where design components are reused. Yet there are opportunities for model building effort and expertise to be more extensively used and reused even within companies involved in small unique projects. The risks, resources and time-scales should be considered in formulating implementation strategy.
6 *Developing in-house competencies and skills* – actively managing use across different functions within the organization, using it to informate rather than automate

processes; considering the resources required and potential benefits. Organizational structures are shifting as new technologies for design visualization begin to be used in project-based firms. Developing in-house competencies in managing design visualization is becoming a key issue. Companies face the real danger that innovation ghettos may be created and that virtual reality will be used to automate rather than informate processes. In the new product development process, experimentation should be managed, with the switching between different media or modes of experimentation optimized to reduce total product development cost and time (Thomke, 1998a). Access to virtual reality should not be confined to one department or project. Even if models are created externally, their use should be actively managed across different functions within the organization.

7 *Working with software suppliers* – to develop next generation solutions and gain competitive edge. Working with software suppliers allows lead users to make use of new technological developments and to shape the next generation of applications. The technologies on which virtual reality is based are in a state of flux and interactive, spatial, real-time applications are being developed and refined in response to feedback from major customers. By working with suppliers, we can make our experience available in commercial tools for the sector.

This checklist is designed as a starting point for organizations implementing virtual reality. Of course, it is crude and partial and may be refined and changed by future experience. Engineering, design and construction organizations are learning by using 3D, interactive 3D and virtual reality. There are still many unanswered questions and this book should be seen as an initial investigation into the subject and a springboard to further work.

References

Where an online version of a reference has been located, a Website address has been included for readers' convenience. Please see the book's homepage http://www.buildingvr.com for a more up-to-date list of online references.

Aouad, G., Child, T., Marir, F. and Brandon, P. (1997). *Open Systems for Construction (OSCON).* Final Report (DOE Funded Project). Salford: University of Salford. http://www.scpm.salford.ac.uk/siene/osconpdf.pdf

Arnheim, R. (1954). *Art and Visual Perception.* Faber.

Baudrillard, J. (1983). *Simulations.* Semiotext(e), Inc.

Bell, G., Parisi, A. and Pesce, M. (1995). *The Virtual Reality Modeling Language Version 1.0 Specification.* http://www.vrml.org/VRML1.0/vrml10c.html

Benedikt, M. (1991). *Cyberspace: first steps.* MIT Press.

Benjamin, W. (1935). *The Work of Art in the Age of Mechanical Reproduction.* http://www.postdogmatist.com/museum/collage/benjamin.htm

Berger, J. (1972). *Ways of Seeing.* Pelican.

Bonsang, K. and Fischer, M. (2000). Feasibility study of 4D CAD in commercial construction. *Journal of Construction Engineering and Management*, 126 (4), 251–60.

Borden, I. (2001). *Skateboarding, Space and the City.* Berg Publishers.

Borges, J.L. (1946). Del rigor en la ciencia. *Los Anales do Buenos Aires.* Credited as 'Suárez Miranda: "Viajes de Varones Prudentes", libro cuarto, cap. XIV, Lérida, 1658'. English translation: Borges, J.L. (1998). On Exactitude in Science. In A. Hurley (trans), *Collected Fictions.* Viking.

Bosselmann, P. (1999). *Representation of Places: reality and realism in city design.* University of California Press.

Bourdakis, V. (1997). The future of VRML on large urban models. *Proceedings of Virtual Reality Special Interest Group VR-SIG'97*, Brunel, UK, 1 November, pp. 55–61. http://fos.prd.uth.gr/vas/papers/UKVRSIG97/index.html

Boyd Davis, S., Huxor, A. and Lansdown, J. (1996). *The Design of Virtual Environments with Particular Reference to VRML*. Report for the Advisory Group on Computer Graphics. http://www.man.ac.uk/MVC/SIMA/vrml_design/title.html

Brooks, F.P. (1986). Walkthrough – a dynamic graphics system for simulating virtual buildings. In S. Pizer and F. Crow (editors), *Workshop on Interactive 3D Graphics*. Association of Computing Machinery (ACM). University of North Carolina at Chapel Hill.

Brooks, F.P. (1992). *Six Generations of Building Walkthroughs*. Final Technical Report, Walkthrough Project, National Science Foundation.

Brooks, F.P. (1999). What's real about virtual reality? *IEEE Computer Graphics and Applications*, 19 (6), 16–27. http://www.cs.unc.edu/~brooks/WhatsReal.pdf

Brooks, F.P., Ouh-Young, M., Batter, J.J. and Kilpatrick, P.J. (1990). Project GROPE: haptic displays for scientific visualization. *Computer Graphics*, 24 (4) 177–85.

Castells, M. (1989). *The Informational City: information technology, economic restructuring, and the urban-regional process*. Blackwell.

Castells, M. (1996). *The Rise of the Network Society*. Blackwell.

Chen, J.L. and Stanney, K.M. (1999). A theoretical model of wayfinding in virtual environments: proposed strategies for navigational aiding. *Presence: Teleoperators and Virtual Environments*, 8 (6), 671–85.

Chu, K. (1998). Genetic space. Profile 136, Architects in Cyberspace II, *Architectural Design*, 68, 68–73.

Conway, B. (2001). Development of a VR system for assessing wheelchair access. *Launch of 4th Call and EQUAL Research Network*, 13 November, Birmingham. http://www.equal.ac.uk/Launch_Posters/equalslidesconway1/sld001.htm

Darken, R.P. and Sibert, J.L. (1993). A toolset for navigation in virtual environments. *Proceedings of the ACM Symposium on User Interface Software and Technology*, Atlanta GA, USA, 3–5 November, pp. 157–65.

Derbyshire, A. (2001). Editorial: probe in the UK context. *Building Research and Information*, 29 (2), 79–84.

Drascic, D. and Milgram, P. (1996). Perceptual issues in augmented reality. *SPIE Volume 2653: Stereoscopic Displays and Virtual Reality Systems III*, San Jose, California, pp. 123–34. http://gypsy.rose.utoronto.ca/people/david_dir/SPIE96/SPIE96.full.html

Dubery, F. and Willats, J. (1972). *Perspective and Other Drawing Systems*. Van Nostrand Reinhold.

Dyson, F. (1998). 'Space', 'being', and other fictions in the domain of the virtual. In J. Beckmann (editor), *The Virtual Dimension: architecture, representation and crash culture*. Princeton Architectural Press, pp. 26–45.

Earnshaw, R.A., Gigante, M.A. and Jones, H. (1993). *Virtual Reality Systems*. Academic Press.

Edgar, G.K. and Bex, P.J. (1995). Vision and displays. In K. Carr and R. England (editors), *Simulated and Virtual Realities: elements of perception*. Taylor & Francis, pp. 85–101.

Edwards, J.D.M. and Hand, C. (1997). MaPS: movement and planning support for navigation in an immersive VRML browser. *VRML 97: Second Symposium on the Virtual Reality Modeling Language*, Monterey, California, pp. 65–74. http://www.cms.dmu.ac.uk/~cph/Publications/VRML97/maps.pdf

English, W., Engelbart, D. and Berman, M. (1967). Display-selection techniques for text manipulation. *IEEE Transactions on Human Factors in Electronics*, HFE-8, 21–31.

Ennis, G., Lindsay, M. and Grant, M. (1999). VRML possibilities: the evolution of the Glasgow model. *VSMM '99*, Dundee, Scotland. http://iris.abacus.strath.ac.uk/gary/papers/glasgow-model.htm

Erickson, T. (1995). Notes on design practice: stories and prototypes as catalysts for communication. In J. Caroll (editor), *Scenario-based Design: envisioning work and technology in system development*. John Wiley & Sons, pp. 37–58. http://www.pliant.org/personal/Tom_Erickson/Stories.html

Feiner, S., MacIntyre, B., Haupt, M. and Solomon, E. (1993). Windows on the world: 2D windows for 3D augmented reality. *Proceedings of the 6th Annual ACM Symposium for User Interface Software and Technology UIST*, Atlanta GA, USA, 3–5 November, pp. 145–55.

Foster, D. and Meech, J.F. (1995). Social dimensions of virtual reality. In K. Carr and R. England (editors), *Simulated and Virtual Realities: elements of perception*. Taylor & Francis, pp. 209–23.

Frampton, R. (2001). *Who Will Buy? An essay accompanying a list of 450+ companies likely to make a significant investment in VR/i3D*. London, Cyber-Wizard Ltd. http://www.VREfresh.com

Frazer, J.H. (1995). The architectural relevance of cyberspace. Profile No. 118, Architects in Cyberspace, *Architectural Design*, 65, 76–79.

Freundschuh, S. (2000). Micro- and macro-scale environments. In R. Kitchin and S. Freundschuh (editors), *Cognitive Mapping: past, present and future*. Routledge, pp. 125–46.

Friedman, T. (1995). Making sense of software: computer games and interactive textuality. In S.G. Jones (editor), *Cybersociety: computer-mediated communication and community*. Sage, pp. 73–89. http://www.duke.edu/~tlove/simcity.htm

Fukuda, T., Nagahama, R. and Nomura, J. (1997). Networked VR system: kitchen layout design for customers. *Proceedings of the 2nd Symposium on Virtual Reality Modeling Language VRML 97*, Monterey CA, USA, 24–26 February, pp. 93–100.

Furness, T.A. (1986). The super cockpit and its human factors challenges. *Proceedings of the 30th Annual Meeting of the Human Factors Society*, Dayton OH, USA, 29 September–3 October, pp. 48–52.

Furness, T.A. (1987). Designing in virtual space. In W.B. Rouse and K.R. Boff (editors), *System Design: behavioral perspectives on designers, tools, and organization*. North-Holland, pp. 127–43.

Gann, D.M. (2000). *Building Innovation: complex constructs in a changing world*. Thomas Telford.

Gann, D.M. and Salter, A.J. (2000). Innovation in project-based, service-enhanced firms: the construction of complex products and systems. *Research Policy*, 29, 955–72.

Gigante, M.A. (1993). Virtual reality: enabling technologies. In R.A. Earnshaw, M.A. Gigante and H. Jones (editors), *Virtual Reality Systems*. Academic Press, pp. 15–25.

Glancey, J. (2001). Fantasy football: Jonathon Glancey builds Leicester City a new stadium in 15 minutes flat. *The Guardian*, 16 May.

Glick, B. (2001). Virtual reality boosts train safety. *Computing*, 19 September. http://www.vnunet.com/News/1125590

Godard, J.-L. (1960). *The little soldier*. France, black & white film, 88 minutes, West End.

Goerger, S.R., Darken, R.P., Boyd, M.A., Gagnon, T.A., Liles, S.W., Sullivan, J.A. and Lawson, J.P. (1998). Spatial knowledge acquisition from maps and virtual environments in complex architectural spaces. *Proceedings of the 16th Applied Behavioral Sciences Symposium*. U.S. Air Force Academy, Colorado Springs CO, USA, 22–23 April, pp. 6–10. http://www.movesinstitute.org/darken/publications/Abss.pdf

Goldin, S.E. and Thorndyke, P.W. (1982). Simulated navigation for spatial knowledge acquisition. *Human Factors*, 24 (4), 457–71.

Gombrich, E.H. (1982). Mirror and map: theories of pictorial representation. In E.H. Gombrich (editor), *The Image and the Eye*. Phaidon, pp. 172–214.

Granstrand, O., Patel, P. and Pavitt, K. (1997). Multi-technology corporations: why they have 'distributed' rather than 'distinctive' core competencies. *California Management Review*, 39 (4), 8–25.

Groák, S. (2001). Representation in building. *Construction Management and Economics*, 19, 249–53.

Gross, M.D. and Do, E.Y.-L. (1996). Ambiguous intentions: a paper-like interface for creative design. *Proceedings of the 9th Annual ACM Symposium for User Interface Software and Technology UIST*, Seattle WA, USA, 6–8 November, pp. 183–92. http://depts.washington.edu/dmgftp/publications/pdfs/uist96-mdg.pdf

Hall, A.C. (1993). Computer visualisation for planning control: objectivity, realism and negotiated outcomes. In M.R. Beheshti and K. Zreik (editors), *Advanced Technologies – architecture, planning, civil engineering*. Elsevier, pp. 303–308.

Hall, P. (1995). Towards a general urban theory. In J. Brotchie, M. Batty, E. Blakely, P. Hall and P. Newton (editors), *Cities in Competition: productive and sustainable cities for the 21st century*. Longman Australia, pp. 3–31.

Harvey, D. (1989). *The Condition of Postmodernity*. Blackwell.

Henderson, K. (1999). *On Line and On Paper: visual representations, visual culture and computer graphics in design engineering*. MIT Press.

Henry, D. (1992). *Spatial perception in virtual environments: evaluating an architectural application*. Master's thesis. HIT Lab. Seattle, University of Washington. http://www.hitl.washington.edu/publications/henry/home.html

Hillier, B. (1996). *Space is the Machine*. Cambridge University Press.

Hobday, M. (1996). *Complex Systems vs Mass Production Industries: a new research agenda*. Working Paper prepared for CENTRIM/SPRU/OU Project on Complex Product Systems, EPSRC, Technology Management Initiative GR/K/31756.

Hockney, D. (2001). *Secret Knowledge: rediscovering the lost techniques of the old masters*. Thames and Hudson.

Ijsselsteijn, W.A., de Ridder, H., Freeman, J. and Avons, S.E. (2000). Presence: concept, determinants and measurement. *Proceedings of the SPIE, Human Vision and Electronic Imaging V*, San Jose CA, USA, 23–28 January, pp. 3959–76.

Isdale, J. (1998). *What is Virtual Reality? A web-based introduction.* http://isdale.com/jerry/VR/WhatIsVR/noframes/WhatIsVR4.1.html

Jacobs, J. (1961). *The Death and Life of Great American Cities*. Vintage.

Jepson, W., Liggett, R. and Friedman, S. (1996). Virtual modeling of urban environments. *Presence: Teleoperators and Virtual Environments*, 5 (1), 72–86.

Jepson, W., Muntz, R. and Friedman, S. (1997). *A Real-time Visualization System for Managing Emergency Response in Large Scale Urban Environments*. White Paper. Los Angeles CA, UCLA. http://www.ust.ucla.edu/~bill/whitepr.html

Johnson, S. (1998). What's in a representation, why do we care, and what does it mean? Examining evidence from psychology. *Automation in Construction*, 9, 15–24.

Johnson, S. (1999). The use of Sim sidewalks. *FEED*.

Johnson, T. (1963). Sketchpad III, three dimensional graphical communication with a digital computer. Master's thesis, MIT. http://theses.mit.edu/Dienst/UI/2.0/Describe/0018.mit.theses/1963-23

Kerr, S. (2000). Application and use of virtual reality. In WS Atkins, *CONVR – Conference on Construction Applications of Virtual Reality: current initiatives and future challenges*. Middlesbrough, UK, 4–5 September, pp. 3–10.

Kiechle, H. (1997). Amorphous constructions 7/97. *morphe:nineteen97 Biennial Oceanic Architecture and Design Student Conference*, Deakin, Australia. http://www.vislab.usyd.edu.au/staff/horst/amorph97.html

Kodama, F. (1995). *Emerging Patterns of Innovation: sources of Japan's technological edge*. Harvard Business School Press.

Krueger, M.W. (1991). *Artificial Reality II*. Addison-Wesley.

Kurmann, D., Elte, N. and Engeli, M. (1997). Real-time modeling with architectural space. *Proceedings of CAAD Futures*, Munich, Germany, 4–6 August, pp. 809–20.

Kwartler, M. (1998). Regulating the good you can't think of. *Urban Design International*, 3 (1), 13–21. http://www.simcenter.org/About_Us/Articles/Reg_Good/reg_good.html

Lawrence, R.J. (1987). House planning: simulation, communication and negotiation. In R.J. Lawrence (editor), *Housing Dwellings and Homes: design theory, research and practice*. John Wiley & Sons, pp. 209–40.

Lawson, B. (2001). *The Language of Space*. Architectural Press.

Lawson, B. and Roberts, S. (1991). Modes and features: the organization of data in CAD supporting the early phases of design. *Design Studies*, 12 (2), 102–108.

Leaman, A. and Bordass, B. (2001). Assessing building performance in use 4: the Probe occupant surveys and their implications. *Building Research and Information*, 29 (2), 129–43.

Lefebvre, H. (1974). La Production de l'Espace. Éditions Anthropos. English edition: Lefebvre, H. (1991) (D. Nicholson-Smith, trans.). *The Production of Space*. Blackwell.

Leigh, J. and Johnson, A.E. (1996). CALVIN: an immersimedia

design environment utilizing heterogeneous perspectives. *Proceedings of IEEE International Conference on Multimedia Computing and Systems '96*, Hiroshima, Japan, 17–21 June, pp. 20–23.
http://evlweb.eecs.uic.edu/spiff/calvin/calvin.mm96/

Ligget, R.S. and Jepson, W.H. (1995). An integrated environment for urban simulation. *Environment and Planning B: Planning and Design*, 22, 291–302.

Lynch, K. (1960). *Image of the City*. MIT Press.

Macheachren, A.M. (1995). *How Maps Work: representation, visualisation and design*. The Guilford Press.

Maher, M.L., Simoff, S. and Gabriel, G.C. (2000a). Participatory design and communication in virtual environments. *PDC2000 Proceedings of the Participatory Conference*, New York NY, USA, 28 November–1 December, pp. 127–134.
http://www.arch.usyd.edu.au/~chris_a/MaherPubs/2000pdf/pdc2000Mark4.pdf

Maher, M.L., Simoff, S., Gu, N. and Lau, K. H. (2000b). Designing virtual architecture. *Proceedings CAADRIA2000*, Singapore, 18–19 May, pp. 481–90.
http://www.arch.usyd.edu.au/~chris_a/MaherPubs/2000pdf/caadria2000.pdf

Marr, D. (1982). *Vision: a computational investigation into the human representation and processing of visual information*. Freeman.

McCullough, M. (1998). *Abstracting Craft: the practised digital hand*. MIT Press.

McGreevy, M.W. (1990). The virtual environment display system. *Proceedings of the 1st Technology 2000 Conference*, Washington DC, USA, pp. 3–9.

McLuhan, M. (1964). *Understanding Media: the extensions of man*. McGraw-Hill.

Mitchell, W.J. (1998). Antitechtonics: the poetics of virtuality. In J. Beckmann (editor), *The Virtual Dimension*. Princeton Architectural Press, pp. 204–17.

Mitchell, W.J. and McCullough, M. (1995). *Digital Design Media*. International Thomson Publishing Inc.

Moss, M.L. and Townsend, A. (1997). *Manhattan Leads the 'Net Nation: New York City and information cities hold lead in Internet domain registration*. New York.

Mumford, L. (1934). *Technics and Civilization*. Harcourt Brace and World, Inc.

Myers, B.A. (1998). A brief history of Human Computer Interaction technology. *ACM Interactions*, 5 (2), 44–54.
http://www-2.cs.cmu.edu/~amulet/papers/uihistory.tr.html

Negroponte, N. (1995). *Being Digital*. Hodder & Stoughton.

Nevey, S. (2001). Oral presentation given at the 3D Design Conference, The Business Design Centre, London.

Novak, M. (1996). Transmitting architecture: the transphysical city. *C-Theory*, 11/29/1996. http://www.ctheory.net/text_file.asp?pick=76

NRC (1999). Virtual reality comes of age. In *Funding a Revolution: government support for computing research*. National Academy Press, Chapter 10. http://www.nap.edu/html/far/

OED (1989). 'virtual, a. (and n.)'. J.A. Simpson and E.S.C. Weiner (editors), *Oxford English Dictionary* 2nd edn, 1989. OED Online. Oxford University Press. 24 October 2001. http://oed.com

Pascucci, E. (1997). Intimate (tele)visions. In S. Harris and D. Berke (editors), *Architecture of the Everyday*. Princeton Architectural Press.

Piaget, J. and Inhelder, B. (1956). *The Child's Conception of Space*. Routledge & Paul.

Pietsch, S.M. (2000). Computer visualisation in the design control of urban environments: a literature review. *Environment and Planning B: Planning and Design*, 27, 521–36.

Pilgrim, M.J., Bouchlaghem, D., Loveday, D. and Holmes, M.J. (2001). A mixed reality system for building form and data representation. *Information Visualisation 2001*, London, UK, 25–27 July, pp. 369–81.

Radford, A., Woodbury, R., Braithwaite, G., Kirkby, S., Sweeting, R. and Huang, E. (1997). Issues of abstraction, accuracy and realism in large scale computer urban models. *Proceedings of CAAD Futures*, Munich, Germany, 4–6 August, pp. 679–90.

Rheingold, H. (1991). *Virtual Reality: exploring the brave new technologies of artificial experience and interactive worlds from cyberspace to teledildonics*. Secker & Warburg.

Richens, P. (2000). Keynote speech: playing games. *International Symposium on Digital Creativity*, Greenwich, UK, 13–15 January.

Robins, K. (1996). *Into the Image: culture and politics in the field of vision*. Routledge.

Robins, K. (1999). Foreclosing on the City? The bad idea of virtual urbanism. In J. Downey and J. McGuigar (editors), *Technocities*. Sage, pp. 34–59.

Roehl, B., Couch, J., Reed-Ballreich, C., Rohaly, T. and Brown, G. (1997). *Late Night VRML 2.0 with Java*. Ziff Davis Press.

Rosenblum, L.J. and Cross, R.A. (1997). Challenges of virtual reality. In R. Earnshaw, J. Vince and H. Jones (editors), *Visualization and Modeling*. Academic Press, pp. 325–39.

Ruddle, R.A., Payne, S.J. and Jones, D.M. (1997). Navigating buildings in 'desk-top' virtual environments: experimental investigations using extended navigational experience. *Journal of Experimental Psychology: Applied*, 3 (2), 143–59.

Satalich, G. (1995). Navigation and wayfinding in virtual reality: finding the proper tools and cues to enhance navigational awareness. Master's thesis, Washington http://www.hitl.washington.edu/publications/satalich/

Sawada, K. (2001). A few recent developments in industrial VR. *Knowledge-Based Intelligent Information Engineering Systems & Allied Technologies (KES'2001)*, Osaka, Japan, 6–8 September, pp. 1609–17.

Sawyer, T. (2001). New York's slow-paced GIS project goes warp speed. *ENR Engineering News Record*, 1 October.

Scaife, M. and Rogers, Y. (1996). External cognition: how do graphical representations work? *International Journal of Human–Computer Studies*, 45, 185–213.

Schön, D.A. (1983). *The Reflective Practitioner.* Harper Collins.

Schrage, M. (2000). *Serious Play: how the world's best companies simulate to innovate*. Harvard Business School.

Schroeder, R., Huxor, A. and Smith, A. (2001). Activeworlds: geography and social interaction in virtual reality, *Futures*, 33 (7), 569–87.

Sherman, W.R. and Craig, A.B. (1995). Literacy in virtual reality: a new medium. *Computer Graphics*, 29 (4), 37–41. http://archive.ncsa.uiuc.edu/VR/VR/Papers/vrlit.html

Shiratuddin, M.F., Yaakub, A.R. and Arif, A.S.C.M. (2000). Utilising first person shooter 3D game engine in developing real world walkthrough virtual reality applications: a research finding. *Conference on Construction Applications of Virtual Reality: Current Initiatives and Future Challenges*, Middlesbrough, UK, 4–5 September, pp. 105–14.

Siegel, A.W. and White, S.H. (1975). The development of spatial representations of large-scale environments. In H. Reese (editor), *Advances in Child Development and Behavior.* Academic Press, pp. 10–55.

Sikiaridi, E. and Vogelaar, F. (2000). *The Use of Space in the Information/Communication Age – processing the unplannable*. Amsterdam, Infodrome. http://www.infodrome.nl/publicaties/domeinen/07_rui_vog_essay.html

Simon, H. (1979). *Models of Thought*. Yale University Press.

Spiller, N. (1998). *Digital Dreams: architecture and the new alchemic technologies*. Ellipsis.

Star, S.L. (1989). The structure of ill-structured solutions: boundary objects and heterogeneous distributed problem solving. In M. Huhs and L. Gasser (editors), *Readings in Distributed Artificial Intelligence 3*. Morgan Kaufmann, pp. 37–54.

Sutherland, I. (1963). Sketchpad, a man–machine graphical communication system. Ph.D. Thesis, MIT.

http://theses.mit.edu/Dienst/UI/2.0/Describe/0018.mit.theses/1963-10
Sutherland, I. (1965). The ultimate display. *Proceedings of the International Federation for Information Processing Societies (IFIPS) Congress*, Vol. 2, pp. 506–508.
Sutherland, I. (1968). A head-mounted three-dimensional display. *Proceedings Fall Joint Computer Conference, American Federation of Information Processing Societies (AFIPS)*. Thompson Book Company. Vol. 33 (1), pp. 757–64.
Suwa, M., Gero, J. and Purcell, T. (1999). Unexpected discoveries: how designers discover hidden features in sketches. *Visual and Spatial Reasoning in Design*, MIT, MA, USA, 15–17 June, pp. 145–62. http://www.arch.usyd.edu.au/~john/publications/PDF2/99SuwaGeroPurcellVR99.pdf
Thomke, S. (1998a). Managing experimentation in the design of new products. *Management Science*, 44 (6), 743–62.
Thomke, S. (1998b). Simulation, learning, and R&D performance: evidence from automotive development. *Research Policy*, 27, 55–74.
Thorndyke, P.W. and Hayes-Roth, B. (1982). Differences in spatial knowledge acquired from maps and navigation. *Cognitive Psychology*, 14, 560–89.
Tufte, E.R. (1997). *Visual Explanations: images and quantities, evidence and narrative*. Graphics Press.
Turner, A., Doxa, M., O'Sullivan, D. and Penn, A. (2001). From isovists to visibility graphs: a methodology for the analysis of architectural space. *Environment and Planning B*, 28 (1), 103–21.
Turner, P. and Turner, S. (1997). Distance estimation minimal virtual environments. *Proceedings of Virtual Reality Special Interest Group VR-SIG'97*, Brunel, UK, 1 November, pp. 90–97.
Vasari, G. (1568). *Le Vite de' più eccellenti pittori, scultori e architettori*. Modern edition in English: Vasari, G. (1998). *The Lives of the Artists* (J.C. Bondanella and P. Bondanella, trans.). Oxford University Press.
Venturi, R., Scott Brown, D. and Izenour, S. (1972). *Learning from Las Vegas*. MIT Press.
von Hippel, E. (1988). *The Sources of Innovation*. Oxford University Press.
von Gerkan, M.P. (2000). *Model Virtuell*. Ernst & Sohn.
Waldrop, M.M. (2000). Computing's Johnny Appleseed. *Technology Review*, January/February. http://www.technologyreview.com/articles/waldrop0100.asp
Watson, K. (2000). Urban design a virtual reality? Master's dissertation, Salford University.
Watts, T., Swann, G.M.P. and Pandit, N.R. (1998). Virtual

reality and innovation potential. *Business Strategy Review*, 9 (3), 45–54.

Whyte, J. (2000). Virtual reality applications in the house-building industry. Ph.D. Thesis, Civil and Building Engineering, Loughborough University.

Whyte, J., Bouchlaghem, N., Thorpe, A. and McCaffer, R. (2000). From CAD to virtual reality: modelling approaches, data exchange and interactive 3D building design tools. *Automation in Construction*, 9 (7), 43–56.

Wickens, C.D. and Baker, P. (1995). Cognitive issues in virtual reality. In W. Barfield and T.A. Furness III (editors), *Virtual Environments and Advanced Interface Design*. Oxford University Press, pp. 514–41.

Witmer, B.G. and Kline, P.B. (1998). Judging perceived and traversed distance in virtual environments. *Presence: Teleoperators and Virtual Environments*, 7 (2), 144–67.

Woods, H. (2000). VRail – virtual reality in the rail environment. *Institution of Railway Signalling Engineers Younger Members Conference, 'The railway as a system'*. Birmingham, UK, 20–21 July.

Wooley, B. (1992). *Virtual Worlds: a journey in hype and hyperreality*. Penguin Books.

Wotton, H. (1624). *The Elements of Architecture ... from the best authors and examples*. John Bill. Modern facsimile reprint: Wotton, H. (1969). *The Elements of Architecture ... from the best authors and examples*. Gregg International Publishers.

Zellner, P. (1999). *Hybrid Architecture: new forms in digital space*. Thames and Hudson.

Zimmerman, A. and Martin, M. (2001). Post-occupancy evaluation: benefits and barriers. *Building Research and Information*, 29 (2), 168–74.

Zuboff, S. (1988). *In the Age of the Smart Machine: the future of work and power*. Basic Books.

Index